T0348438

New Carbons

Control of Structure and Functions

New Carbons

Control of Structure and Functions

MICHIO INAGAKI
Aichi Institute of Technology
Department of Applied Chemistry
Yakusa, Toyota 470-0392, Japan

2000

ELSEVIER
Amsterdam – Lausanne – New York – Oxford – Shannon – Singapore – Tokyo

ELSEVIER SCIENCE Ltd.
The Boulevard, Langford Lane
Kidlington, Oxford OX5 1GB, UK

First edition 2000

Library of Congress Cataloging in Publication Data
A catalog record from the Library of Congress has been applied for.

British Library Cataloguing in Publication Data
A catalogue record from the British Library has been applied for.

ISBN: 0 08 043713 3

♾ The paper used in this publication meets the requirements of ANSI/NISO Z39.48-1992 (Permanence of Paper).
Transferred to digital printing 2005
Printed and bound by Antony Rowe Ltd, Eastbourne

Preface

The discovery of fullerenes and nanotubes has greatly stimulated the interest of scientists and engineers in carbon materials, and has resulted in much scientific research. These materials have provided us with many interesting ideas and potential applications, some of them practical and some simply dreams for the future. Carbon materials have a long history starting from charcoal used in prehistoric times. The development of graphite electrodes for steel refining supported the revolution in industry and greatly contributed to modern developments in various industries; carbon blacks have been used in black inks and also as reinforcing fillers for various tyres used in the present motor industry. Activated carbons have wide applications which help maintain the quality of our lives nowadays, e.g. purifying water and deodorizing air, which have become increasingly important.

In the early 1960's, carbon fibres, glass-like carbons and pyrolytic carbons were developed which were quite different from the carbon materials that had previously been used. Carbon fibres exhibited surprisingly good mechanical properties, glass-like carbons exhibited brittle fracture resulting in a conchoidal fracture surface similar to sodium glass, and giving no carbon dust, and pyrolytic carbons were produced by a new production process of chemical vapour deposition and showed very high anisotropy. These carbon materials made a great impact not only on the carbon community who had been working on carbon materials but also on people working in the fields of materials science and engineering. They were used to develop a variety of new applications in technological fields, such as semiconductors, microelectronics, aerospace and high temperature, etc. These newly developed carbon materials were called NEW CARBONS, in comparison with carbon materials such as artificial graphites represented by graphite electrodes, carbon blacks and activated carbons, which may be thought of as CLASSICAL CARBONS. Later, other new carbons, such as activated carbons and those with novel functions, isotropic high-density graphites, intercalation compounds, various composites, etc., were developed.

In 1994, the author published a book entitled "New Carbon Materials – Structure and Functions" with his friend Prof. Yoshihiro Hishiyama of Musashi Institute of Technology, published by Gihoudou Shuppan in Japanese. However, progress in the fields of these new carbons is so rapid that the previous book is already out of date. Also numerous developments related to new carbons have been made in Japan, but regretfully some of them have not been published in English and they are therefore unknown by the international carbon community. For these reasons the author has decided to write an English text on New Carbons.

In this book, the author has focused on New Carbons based on hexagonal networks of carbon atoms, in other words, graphite-related materials. Diamonds, diamond-like carbons, fullerenes and nanotubes are not included. One of the reasons for this is that even though they may be classified as New Carbons there are already many specialised books about them. In addition, it is the graphite-related materials which have been the focus of the author's interests, and he believes there has been pronounced progress in such New Carbons in recent years.

The fundamental concept of the present book is that the structure and functions of graphite-related carbon materials are principally governed by their texture, *i.e.*, the mode and degree of preferred orientation of anisotropic hexagonal carbon layers, for which a means of classification and definition has been proposed by the author. After giving an overview of carbon materials, including allotropes, and a short description of fundamental characterisation techniques for carbon materials, the structure and functions of new carbons are explained in 6 chapters, which have been divided based on their textures, both micro and macro.

The author hopes to give readers a comprehensive understanding of New Carbon Materials through the description of their structure and texture, and also of properties that are largely dependent upon them. He does not review all related published papers, even though he might want to, because not only is this almost impossible but there is also the fear that he might not correctly reproduce the ideas and results of some of the papers referred to. He would therefore very much appreciate receiving frank comments on this book, both critical and constructive, from its readers.

February, 2000

Michio Inagaki

Contents

CHAPTER 1

Old but New Materials: "Carbons"

1.1. Carbon Materials

Carbon is a very important element for all living things on the Earth, including humans, because all organic compounds are composed from carbon networks. Carbon materials, which consist mainly of carbon atoms, have been used since prehistoric times in the form of charcoal. In Japan, a large amount of charcoal (about 800 tons) was reported to be used for casting a great image of Buddha in Nara from 747 to 750. Diamond crystals, which are fascinating not only as jewels but also as the hardest known material, were found to consist of carbon atoms, the same atoms as lubricating soft graphite, as recently as 1799. Soft graphite has been used for a long time as pencil lead and carbon blacks as black ink. Graphite electrodes, which were essential for metal refining, are still used and produced in large amounts (in 1975 a maximum production of approximately 800,000 tons in Japan was recorded) and Fig. 1-1 shows large (700 mm in diameter and 2.8 m long) electrodes with nipples used for iron refining. Carbon blacks (Fig. 1-2) of different sizes have also been used in large amounts (approximately 30,000 tons in Japan in 1982), small ones for tires and large ones for wet suits. Figure 1-3 shows activated carbon produced from coconut shells for a tobacco filter, the production of which has been increasing annually (approximately 50,000 tons in Japan in 1982). These three carbon materials (electrode graphites, carbon blacks and activated carbons) have been used for a long time and may be called classical carbon materials, in contrast to newly developed carbon materials, which may be called *new carbons*.

Various carbon materials are now used in various aspects of human activity. Many of them are inconspicuous, e.g. carbon fibers for the reinforcement of rackets and fishing rods, activated carbons as filters for deodorization in refrigerators and in water purification to produce Japanese sake, membrane switches for the keyboards of computers and other equipments, and lead for automatic pencils. Some examples of newly developed carbon materials, some of the new carbons, are shown in Figs 1-4 to 1-6. The speaker cone shown in Fig. 1-4a is made of a carbon/resin composite, which gives a good-quality sound in the bass region. Springs made of carbon (Fig. 1-4b) have higher thermal resistance than metallic ones. Leads for sharp pencils (Fig. 1-5) and membrane switches for various electronics (Fig. 1-6) are now commonly used in our daily lives. Lithium rechargeable batteries using carbon anodes have made possible portable electronic devices, mobile computers and portable telephones, which are used world-wide.

Carbons are strange and interesting materials because topics on some carbon materials have arisen roughly every 5 years. Table 1-1 summarizes the developments since 1960 by dividing into them fundamental science, materials development and technology related to applications. The year 1960 may be thought of as the beginning of the era of *new carbons*, because of the invention of carbon fibers from polyacrylonitrile, pyrolytic carbons and glass-like carbons.

People have been fascinated by the high strength and flexibility of carbon fibers, and many demonstrative pictures, e.g. hanging a car by thin strings of carbon fiber, have been published in various journals. The development of other kinds of carbon fiber, i.e. mesophase-pitch-based and vapor-grown ones, was followed around the 1970s. Today, these types of carbon fiber are produced on an industrial scale and used in various fields (Chapter 4 in this book).

In contrast, glassy carbons (glass-like carbons) represent hard carbons, showing conchoidal fracture surfaces as sodium glass, from which the name "glassy carbon" originates. Its gas impermeability had never been previously realized (Chapter 3).

Pyrolytic carbons were produced by chemical vapor deposition (CVD), which was a completely different technique from conventional ones, although it has since become very common in material production. Their strong anisotropy in various properties provided a

Figure 1-1 Large graphite electrodes with graphite nipples, used for iron refining. [Courtesy of Tokai Carbon Co. Ltd.]

quite new aspect for the application of carbon materials, such as highly oriented pyrolytic graphite (HOPG) as a monochromator for X-rays (Chapter 2).

In 1964, the formation of optically anisotropic spheres in pitches (mesophase spheres) and their coalescence were demonstrated. The following detailed studies on these mesophase spheres, the structure of the spheres, growth and coalescence of spheres and formation of bulk mesophase, created needle-like cokes which were the essential raw materials for high-power graphite electrodes, mesophase-pitch-based carbon fibers with high performance and mesocarbon microbeads for different applications.

Around 1970, carbon materials were found to have good biocompatibility and various prostheses, such as heart valves and tooth roots, were developed. Around 1980, industrial technology for producing isotropic high-density graphites (Chapter 3) using rubber-presses was established, creating applications for high-temperature gas-cooled reactors, the synthesis of semiconductor crystals and electric discharge machining. Around 1985, the mixing of a small amount of carbon fibers into cement paste was found to result in a pronounced reinforcement of concrete. Today, not only carbon fiber reinforced concrete but also carbon fibers themselves are used in the field of civil engineering, such as in buildings, bridges and various other constructions (Chapters 4 and 7).

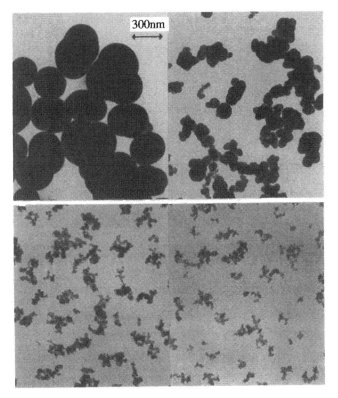

Figure 1-2 Transmission electron micrographs of carbon blacks of different sizes. [Courtesy of Tokai Carbon Co. Ltd.]

The high electrical conductivity of the AsF$_5$-graphite intercalation compound, which is higher than that of metallic copper, had a strong impact on scientists and engineers, which induced a boom in research on graphite intercalation compounds. This research did not lead to practical applications of these intercalation compounds, mainly because of their poor stability. However, research is continuing in order to develop a practical use for carbon materials as anode materials for lithium ion rechargeable batteries (Chapter 6).

The discovery and synthesis of buckminsterfullerene C$_{60}$ and the superconductivity of its potassium compound K$_3$C$_{60}$ opened a new chapter in carbon materials, as will be explained in the next section, and created a new world-wide boom. Large fullerenes, such as C$_{70}$ and C$_{76}$, some giant fullerenes, such as C$_{540}$, multiwall fullerenes, single-wall nanotubes and multiwall nanotubes followed.

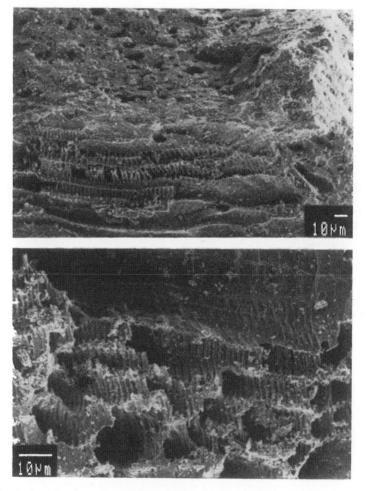

Figure 1-3 Activated carbon prepared from coconut shells. [Courtesy of Prof. Y. Hishiyama and Dr A. Yoshida of Musashi Institute of Technology.]

Figure 1-4   Speaker cone and springs made of carbon. [Courtesy of Mitsubishi Pencil Co. Ltd.]

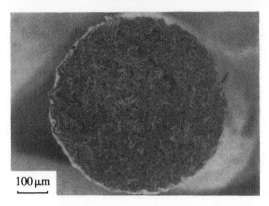

100 μm

Figure 1-5   Scanning electron micrographs of a lead for a sharp pencil. [Courtesy of Prof. Y. Hishiyama and Dr A. Yoshida of Musashi Institute of Technology.]

1.2. Carbon Family

It is well known that carbon atoms can have three different hybrid orbitals, sp^3, sp^2 and sp, giving a variety of chemical bonds. This variety makes possible an enormous number of hydrocarbons, from which a great number of organic materials can be constructed. The C–C bond using sp^3 and sp^2 hybrid orbitals was known in the construction of diamond and graphite, respectively, in carbon materials, which are inorganic materials. However, the recent discoveries of buckminsterfullerene C_{60} and carbyne, which is supposed to be constructed from sp bonds, may renew the importance of the bonding nature between carbon atoms. Figure 1-7 illustrates how the variety in the bonding nature of carbon atoms leads to a large family of organic molecules and that the inorganic carbon materials, diamond, graphite, fullerenes and carbyne, have resulted from extensions to become giant molecules of these organic materials.

Here, a family of inorganic carbon materials, the carbon family, may be defined, of which diamond, graphite, fullerene and carbyne are members. The structural character-istics and structural variety in each family are summarized in Fig. 1-8.

Diamond consists of sp^3 orbitals, where chemical bonds extend in three-dimensional space and are purely covalent. It is very hard because of its covalent bonds and is an electrical insulator because of the high localization of electrons. Long-range periodical repetition of this bond forms diamond crystal. Most diamond crystals belong to the cubic system, but some to the hexagonal system, which can be understood by its resemblance to the zincblende and wurtzite-type structures, respectively, in ZnS and BN. In the case where long-range periodicity is not attained either by the introduction of defects or by hydrogen but sp^3 bonding is kept, diamond-like-carbon (DLC) with an amorphous structure is formed.

The carbon family with sp^2 bonding is represented by graphite, where the layers of hexagons of carbon atoms bound using sp^2 orbitals are stacked in parallel using π electron clouds with a regularity of ABAB..., which belongs to the hexagonal crystal system. A stacking regularity of ABCABC... is also possible, which belongs to the rhombohedral crystal system, but it occurs only locally by introducing stacking faults due to shearing stress during grinding, for example. The parallel stacking of the layers with complete randomness can be found mostly in the carbon materials prepared at low temperatures, such as 1300°C, where the layers of hexagons are usually small and a few layers are stacked in parallel. This structure, consisting of random stacking of layers, is called a turbostratic structure [Warren, 1934]. Regular and random stacking, i.e. graphitic and turbostratic, respectively, are illustrated in Fig. 1-9. By heating these carbons to high temperatures, up to 3000°C, both the size and number of stacked layers usually increase and the regularity of stacking is improved. In other words, a wide range of structure exists, from completely turbostratic stacking to pure ABAB stacking, through intermediates with variable ratios of these two types of stacking, which depend primarily on the starting materials (precursors) and heat treatment temperature (HTT) and, as a consequence, a wide variety of properties [Inagaki and Hishiyama, 1999]. The carbon materials belonging to this carbon family, based on a graphitic structure, are electrically and thermally

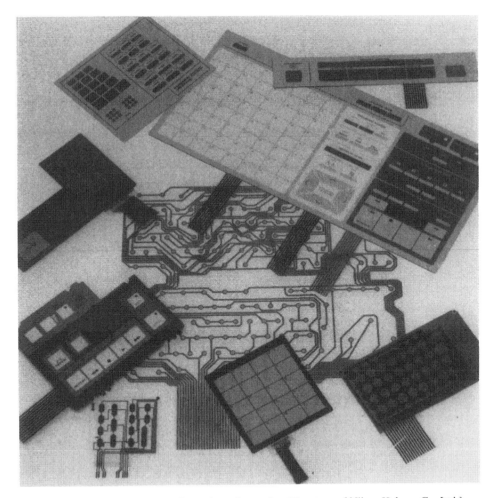

Figure 1-6   Membrane switches for various electronics. [Courtesy of Nihon Kokuen Co. Ltd.]

conductive and soft, mainly because of the presence of π electrons, in sharp contrast to diamond.

In this family, the fundamental structural unit is a layer of carbon hexagons (sometimes called graphene) and, as a consequence, the materials have a strong anisotropy because the bonds in the layer are localized to be covalent and those between the layers van der Waals like. Therefore, the way in which these layers are agglomerated, in addition to the mixing ratio of ABAB and turbostratic stackings, leads to a variety of materials. A classification based on the scheme and degree of preferred orientation of anisotropic layers has been proposed [Inagaki, 1985a] and is used successfully (Fig. 1-10).

Initially, random and oriented textures have to be differentiated and then the latter are classified by the scheme of orientation along a reference plane (planar orientation), along the reference axis (axial orientation) and around a reference point (point orientation). The extreme cases of these orientation schemes are a single crystal of graphite for planar orientation, multiwall nanotubes for axial orientation and multiwall fullerenes for point orientation. A texture with random orientation occurs in glass-like carbons. Various

Table 1-1. Topics related to carbon materials

Year	Basic science	Materials development	Technological development
1960		PAN-based carbon fibers Pyrolytic carbons Glass-like carbons	Electrode for electric discharge machining
1965	Mesophase spheres	Needle-like cokes Mesophase-pitch-based carbon fibers	
1970	Biocompatibility of carbons	Vapor-grown carbon fibers	Carbon prostheses
1975	High conductivity of GICs	Isotropic high-density graphites	Mesocarbon microbeads
1980			
	i-carbon films	Carbon fiber reinforced concrete	Carbon electrode for fuel cell
1985	Buckminster-fullerene C_{60}		
			First wall for fusion reactor
1990	Superconductivity of K_3C_{60}		
1995			Carbon anode for lithium ion rechargeable batteries

PAN, polyacrylonitrile; GIC, graphite intercalation compound.

intermediates in planar orientation are found in pyrolytic carbons depending on HTT. In axial orientation, concentric and radial alignments of layers relative to the reference axis

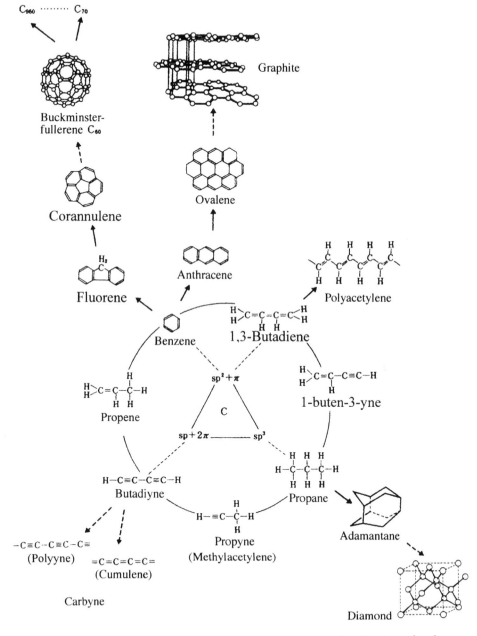

Figure 1-7 Enormous organic compounds based on carbon–carbon bonds using sp³, sp² and sp hybrid orbitals and inorganic materials as their extension.

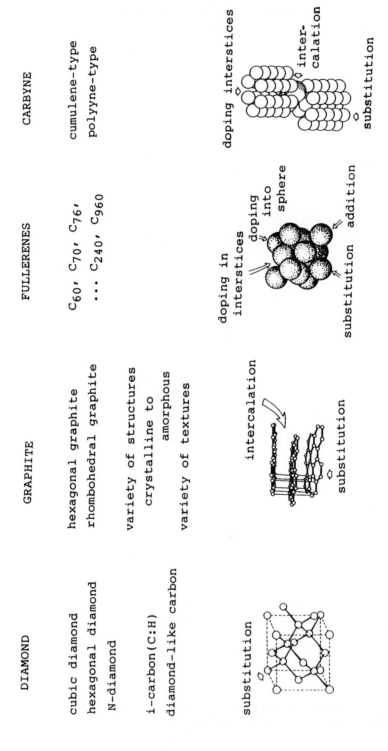

Figure 1-8 Structural characteristics and possibilities to develop the function of the carbon family.

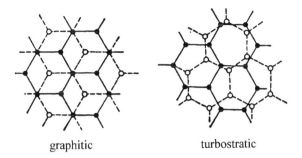

graphitic   turbostratic

Figure 1-9 Scheme of graphitic and turbostratic stacking of carbon layers.

are possible, the former being in vapor-grown carbon fibers and the latter in one of mesophase-pitch-based carbon fibers. In point orientation, concentric and radial alignments also have to be differentiated: the former can be found in carbon black particles and the latter in the spheres formed from a mixture of polyethylene and polyvinylchloride by pressure carbonization [Inagaki, et al., 1981; Hishiyama et al., 1982]. The structure of

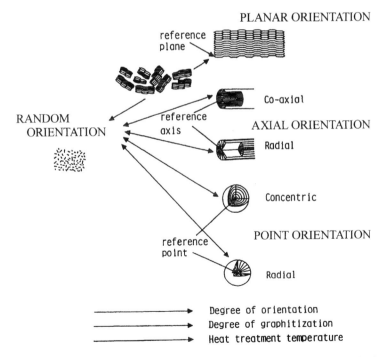

Figure 1-10 Texture on a nanometric scale in carbon materials based on the preferred orientation of anisotropic hexagonal layers [Inagaki, 1985c].

mesophase spheres is close to the radial point orientation scheme, but in their centers the orientation of layers is not radial [Brooks and Taylor, 1968; Auguie et al., 1980]. In these spherical particles mentioned above, hexagonal carbon layers are aligned radially on the surface by making parallel of latitude. Recently, however, spheres with a radial point orientation texture and a random arrangement of hexagonal layers at the surface were obtained by a combustion flame method [Despres et al., 1997].

Most particles with planar and axial orientation, e.g. pyrolytic carbons and carbon fibers, are also anisotropic and so their agglomeration can create further variety in texture. Therefore, it is necessary to take into consideration the scale of constituent units of the texture: texture due to the preferred orientation of anisotropic hexagonal layers is on a nanometric scale (Fig. 1-10), composites based on the piles of the clothes of carbon fibers are composed of the texture on a micrometric scale and graphite electrodes have the texture on a millimetric scale of coke grains with carbon formed from binder pitches. Therefore, the variety of structure in this graphite-based family is mainly due to the size of hexagon layers, their stacking order and their orientation scheme.

The nature of bonding in fullerene particles is expected to be close to the sp^2 bond, but this is not yet clear. The particle of buckminsterfullerene C_{60} is composed of 12 pentagons and 20 hexagons of carbon atoms. The increase in the number of hexagons in C_{60} to make all pentagons apart from each other leads to giant fullerenes, and that to make two groups of six pentagons apart from each other results in nanotubes (Fig. 1-11). In this carbon family, the variety of structure is mainly due to the number of carbon atoms consisting of fullerene particles and the relative location of 12 pentagons.

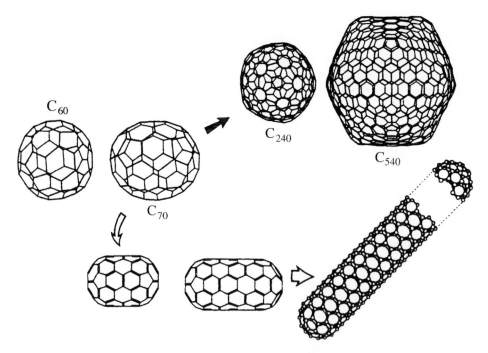

C_{60}

C_{70}

C_{240}

C_{540}

Figure 1-11   Expansion from buckminsterfullerene C_{60} to giant fullerene C_{540} and a nanotube.

Carbyne has been supposed to be carbon atoms bound linearly by sp bonding [Kudryavtsev et al., 1997], where two π electrons have to be resonated, giving two possibilities, an alternative repetition of single and triple bonds (cumulene) and a simple repetition of double bonds (polyyne). Its detailed structure has not yet been clarified, but one proposed structural model is illustrated in Fig. 1-8, where some carbon atoms form a line using the sp hybrid orbital, these lines gather by van der Waals interaction between π electron clouds to make a layer, and then these layers are stacked. In this carbyne family, the variety of structure is mainly due to the number of carbon atoms forming a line, in other words, the thickness of layers of linear carbon chains, and the density of chains in a layer.

In these carbon families, doping or insertion of foreign atoms leads to a further expansion of functions. Each family has different possibilities, as summarized in Fig. 1-8. In the diamond family only the substitution of foreign atoms for carbon atoms is possible, but in the graphite family intercalation of a number of species (ions and molecules) can occur (intercalation compounds; Chapter 6), in addition to the substitution of foreign atoms. In fullerenes, four possibilities can be mentioned: doping in the interstitial sites of fullerene particles, doping into the inside of particles, substitution of foreign atoms for carbon atoms in the particle, and adduction of some organic radicals on to the surface of the particle. In the carbyne family, intercalation between layers of carbon chains, doping into the space between carbon chains in a layer and substitution are considered. The intercalation of either iron or potassium atoms has been reported to stabilize the carbyne structure.

1.3. Purpose of the Present Book

The present book focuses on graphite-based carbon materials, particularly the "*new carbons*". The author hopes to present a comprehensive understanding on carbon materials through the description of their crystallographic structure and texture, and also on various functions that largely depend on the structure and texture.

It should be mentioned here that the author does not intend to review all published papers related to the carbon materials described, but rather to give an overview on each new carbon material on the basis of their structure and texture, which he has seen to be the most important factors in understanding carbon materials during his 40 years of research activity. Therefore, it should be stated in advance that the references are not complete and also that he prefers to refer to works and papers published by Japanese research groups, because some of them have not been published in English and are not yet known in the rest of the world. He regrets that they have not been published in English, even though they are very interesting and important in creating the future of carbon materials. It will be a great pleasure for the author if readers understand his point of view and his fundamental concept for carbon materials.

The author paid particular attention to the crystallographic structure and texture of new carbons, because these factors strongly govern their functions. In relation to the discussion on their crystallographic structure and texture, their production procedures have had to be explained briefly, but not in detail. Therefore, readers who require details of the production and preparation procedures should refer to the original papers and appropriate published reviews. In addition, the fundamental theory, equipment and procedures used for the characterization techniques are not detailed in this book, except for some important points needed to study carbon materials, on X-ray powder diffraction, magnetoresistance, transmission electron microscopy and scanning electron microscopy. Readers are also expected to refer to books specializing in each technique.

1.4. Characterization Techniques for Carbon Materials

Characterization, particularly of the structure and texture of carbon materials, is essential and important in order to use them correctly, as well as to understand them. Before going on to discuss each new carbon material in the following chapters, fundamental techniques for the characterization of the structure and texture of carbon materials are explained here, particularly emphasizing the special care required in applying them to carbon materials. However, books specializing on each technique should be consulted for details of the principles and equipment.

(a) X-ray powder diffraction

X-ray diffraction is one of the fundamental techniques used to characterize crystal structure. A powder diffractometer with a counter is usually employed to study carbon materials because most of them are polycrystalline. However, special attention has to be paid to carbon materials, and particularly anisotropic carbon materials.

Figure 1-12a and b shows X-ray diffraction patterns of the powders of natural graphite and low-temperature-treated coke, respectively. The diffraction peaks observed on natural graphite in Fig. 1-12a are sharp and coincide well with those expected from an ideal graphite structure in *hkl* indices and the interlayer spacing of the peaks. However, the relative intensities of the diffraction lines are different from those expected theoretically, usually *00l* lines (*002*, *004* and *006* in Fig. 1-12a), having a much higher intensity, and *hk0* lines a much lower intensity than expected. This is due to the preferred orientation of graphite flakes in a flat-plate sample holder for the diffractometer. From geometry among the X-ray source, sample and counter in this diffractometer, the counter can record only the diffracted beam from the crystal planes, which are parallel to the surface of the flat-plate sample holder. In the case of natural graphite powder, the flakes tend to orient making their surfaces, which consist of graphite layers, parallel to the holder surface during their packing and, as a consequence, the second, fourth and sixth orders of diffraction lines due to graphite layers, i.e. *002*, *004* and *006*, are preferentially observed.

A typical example is shown in Fig. 1-13; the sample is a thin flake of HOPG, in which graphite layers are highly oriented along the flake, and diffraction patterns are taken by reflection mode, as in the case of natural graphite in Fig. 1-12a, and by transmission mode, a schematic illustration on the relation between X-ray path directions and crystallographic planes being shown in the figure. Only the diffraction lines due to *00l* planes are observed in reflection mode and those due to *hk0* planes, which are due to the crystal planes perpendicular to the layer planes, in transmission mode.

The powder pattern for low-temperature coke in Fig. 1-12b is quite different from that of natural graphite. Two differences have to be pointed out: the appearance of two-dimensional *hk* lines, instead of *hkl* lines in natural graphite, and broad and shifted *00l* lines. The appearance of *hk* lines is due to turbostratic stacking of hexagonal layers, lacking a regularity along the *c*-axis (therefore, no *l* index) and their characteristic asymmetric profiles have been explained. By applying Fourier analysis on this asymmetric *hk* line, the probability P_1 of finding graphitic AB stacking of layers is calculated [Warren, 1934]. This probability P_1 indicates how many hexagonal layers have a graphitic relation with neighboring layers, and so can be called the degree of graphitization. Through Fourier analysis, the probabilities of having ABA and ABC stacking (occurring in hexagonal and rhombohedral graphites, respectively), P_{ABA} and P_{ABC}, can also be calculated.

00l lines in low-temperature coke are broad and shifted to the low diffraction angle side. Since the position of the diffraction line, diffraction angle 2θ, gives an averaged value of interlayer spacing, this coke has much higher average spacing between neighboring two hexagonal layers, d_{002}, than graphite (0.3354 nm). In most carbon materials, the average interlayer spacing d_{002} is usually larger than that of graphite and approaches it gradually through high-temperature treatment. This is explained by the model of random mixing of graphitic and turbostratic stacking in a structural unit, as shown schematically in Fig. 1-14a. This structural unit has been called "crystallite" even though it is not exactly correct terminology according to crystallography because of the presence of random (turbostratic) stacking in it. Their averaged thickness and diameter, L_c and L_a, are determined from *00l* and *hk0* diffraction lines, respectively, in the X-ray powder pattern. The values of L_c

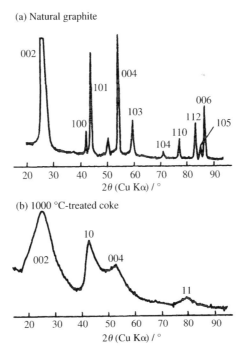

Figure 1-12　X-ray diffraction patterns of natural graphite (a) and low-temperature-treated coke (b).

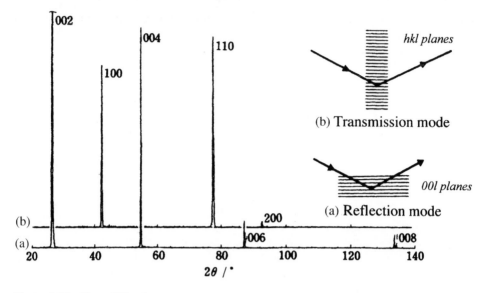

Figure 1-13 X-ray diffraction patterns measured by reflection mode (a) and transmission mode (b) on highly oriented pyrolytic graphite (HOPG) [Inagaki and Hishiyama, 1994c].

determined from the half-width of *002*, *004* and *006*, if possible, are in most cases different, which can be explained by the presence of random stacking in the crystallites [Inagaki, 1968; Murase et al., 1968]. Therefore, it has been recommended to show the

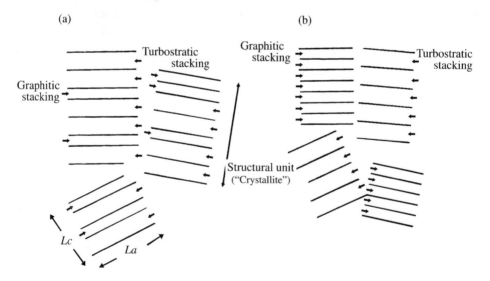

Figure 1-14 Schematic illustration of the random formation of graphitic stacking in a crystallite (a) and that of the coexistence of two crystallites with graphitic and turbostratic stacking (b).

index of the diffraction line used, e.g. $L_c(002)$ and $L_c(004)$. The spacing in turbostratic stacking is reasonably supposed to be larger than that in regular graphitic stacking. The spacing of 0.344 nm has been assigned to it [Franklin, 1951], but even larger spacing has been found in various carbon materials.

If these two different stacking sequences, graphitic and turbostratic, occur in separated crystallites, as in Fig. 1-14b, the 002 profile has to become composite, overlapping two peaks with different 2θ positions. This happens in the powder mixtures and composites, not only a mixture of natural graphite with low-temperature coke but also carbon fiber/ carbon composite. The broadening of diffraction lines is due to the small size of crystallites and the presence of strain in the structure. By ignoring the structural strain, the crystallite size can be calculated from the half-maximum width β (line-width in 2θ at half-intensity of the line) of each line by Scherrer's equation. The calculated values from 002 and 004 lines of carbon materials, however, are usually different, as discussed above. By assuming the presence of different interlayer spacings, as is the case in the model of Fig. 1-14a and which may be expressed by the term "strain", the following relation was proposed [Inagaki et al., 1973; Inagaki, 1985a]:

$$\beta_{obs} \cos \theta/\lambda = K/L + 2\varepsilon \sin \theta/\lambda,$$

where β_{obs} is the observed width of different $00l$ lines, θ is the diffraction angle for $00l$ lines, K is a constant, L is the crystallite thickness along the c-axis, λ is the wavelength of X-ray used and ε is the strain due to the presence of different interlayer spacings.

Carbon has a relatively low absorption coefficient for X-rays; in other words, X-rays penetrate into the sample carbon. The X-ray beam diffracted at the deep position of the sample is recorded at a lower angle than that from the surface, which may broaden the line profile and shift the position of the line to the lower angle side. Therefore, the use of thin samples and inner standards is recommended. Japanese specification for X-ray measurements on carbon materials recommends the following two points: 1) the use of a glass sample holder with a depth for sample powder of 0.2 mm; and 2) mixing of silicon fine powder with high purity as an inner standard into sample powders [Inagaki, 1963b].

In order to express the degree of structural development in carbon materials, the parameters p [Franklin, 1951; Bacon, 1951] and g [Mering and Maire, 1965] were proposed, which could be calculated from d_{002}, assuming that turbostratic stacking had a constant spacing of 0.344 nm. However, this assumption for turbostratic stacking with a constant spacing has not been proved by experimental data, with even much larger spacing than 0.344 nm being observed on various carbon materials. Therefore, the interlayer spacing d_{002}, which is determined from 002 and 004 lines, is commonly used. The probability P_1 is the most exact parameter to evaluate the development of graphite structure in carbon materials. This parameter P_1 was also reported to be uniquely related to d_{002} [Houska and Warren, 1954; Bowman, 1956; Noda et al., 1966], but it was shown in recent studies on a wide range of carbon materials that the relation between P_1 and d_{002} depends strongly on the textures of carbon materials [Iwashita and Inagaki, 1993]. The relations among structural parameters measured by X-ray diffraction, d_{002}, L_c, L_a, ε, etc., have been discussed [Inagaki, 1968; Murase et al., 1968; Iwashita and Inagaki, 1993; Iwashita et al., 1993].

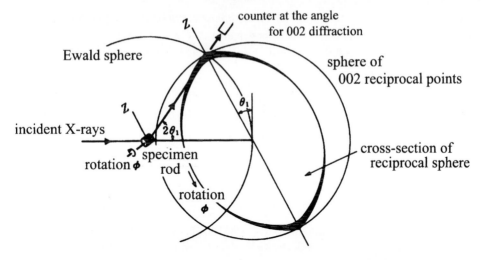

Figure 1-15 Principle for the measurement of orientation function in carbon materials.

The orientation of graphitic layers is also measured by using a diffractometer, the principle being shown by using a reciprocal lattice sphere composed of *002* reciprocal points for each of the layers in Fig. 1-15. The sample must be a thin rod of which the axis must be perpendicular to the orientation direction. By rotating the rod sample, *002* diffraction intensity is measured as a function of rotation angle ϕ, the curve observed being called the orientation function $I(\phi)$. Some orientation functions measured on different carbon materials are shown in Fig. 1-16. On the samples with a well-defined and high degree of orientation, a width at 50% intensity is employed as a measure of orientation $\phi_{1/2}$.

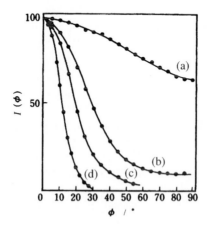

Figure 1-16 Some orientation functions. (a) Gilsonite coke; (b) a needle-like coke; (c) a well-oriented coke prepared under a magnetic field and heat-treated at 3000°C; (d) a pyrolytic carbon heat-treated at 3100°C.

On some sample plates with a relatively high orientation degree, an intensity change in the *002* line could be measured by fixing the counter at the diffraction angle of *002* and rotating the sample plate a little. In this geometry for measurement, however, the sample volume irradiated by the X-ray changes with the rotation of the sample and finally the X-rays are blocked out of the counter. Therefore, the exact orientation function cannot be obtained, but it can be calculated after appropriate correction on this geometry in the case of well-oriented sample. The half-width of the orientation function thus measured is called mosaic spread (MS). However, it has to be noted that this measurement of orientation function must be limited to well-oriented samples.

(b) Magnetoresistance

Transverse magnetoresistance $\Delta\rho/\rho$, is defined as a relative change in electrical resistivity ρ with and without magnetic field B:

$$\Delta\rho/\rho = [\rho(B) - \rho(0)]/\rho(0).$$

It was found to give useful parameters for characterization of carbon materials [Hishiyama et al., 1991]. As seen in the above equation, magnetoresistance is a relative change in electrical resistivity and depends only on electrical current for the measurement of resistivity ρ and magnetic field strength B applied, not influenced by the geometry, size or shape of the specimen. However, it depends on the orientation of crystallites in the specimen because of their strong anisotropy in electrical resistivity. Therefore, the magnetoresistance values of carbon materials are measured by applying the magnetic field in three orthogonal directions, as shown in Fig. 1-17.

On a specimen with a planar orientation of crystallites along the plane of specimen, the direction *max* of the magnetic field which gives a maximum value of $\Delta\rho/\rho$, $(\Delta\rho/\rho)_{max}$, is

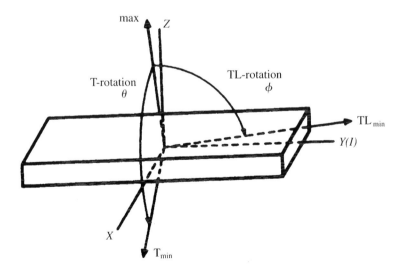

Figure 1-17 Rotation directions for measuring the magnetoresistance of carbon plate [Hishiyama et al., 1991].

first defined and then two orthogonal directions, TL_{min} and T_{min}, the former giving a minimum value of magnetoresistance, $(\Delta\rho/\rho)_{TLmin}$, near the direction of current and the latter being perpendicular to these two directions, $(\Delta\rho/\rho)_{Tmin}$. In Fig. 1-18, the dependences of $\Delta\rho/\rho$ on angle of TL and T rotations, ϕ and θ, are shown on a needle-like coke with different HTT. In each dependence, the maximum absolute value of $\Delta\rho/\rho$ corresponds to $(\Delta\rho/\rho)_{max}$ and two minima for TL and T rotations to $(\Delta\rho/\rho)_{TLmin}$ and $(\Delta\rho/\rho)_{Tmin}$, respectively.

With the increase in HTT, $\Delta\rho/\rho$ changes its sign from negative to positive, indicating the change in charge carriers from positive holes to the coexistence of positive holes and negative electrons. In addition, its rotation angle dependences become pronounced and there is not so much difference in the two rotation directions TL and T, showing that the principal orientation scheme is planar one and the degree of orientation increases markedly with increasing HTT.

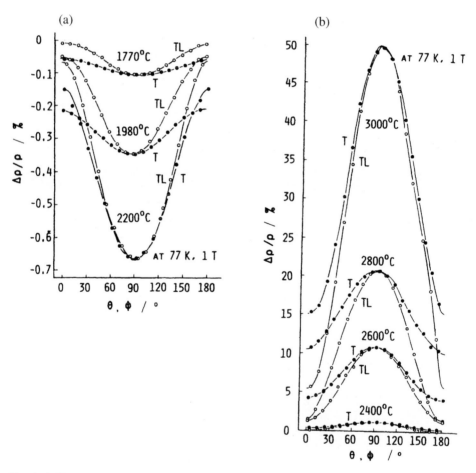

Figure 1-18 Changes in magnetoresistance $\Delta\rho/\rho$ with rotation of the magnetic field along two directions, TL and T, on needle-like coke particles with different HTT [Hishiyama and Kaburagi, 1979].

In order to characterize the structure and texture of carbon materials, the following three parameters, crystallite magnetoresistance $(\Delta\rho/\rho)_{cr}$ and anisotropy ratios r_{TL} and r_T, are defined from these measurements:

$$(\Delta\rho/\rho)_{cr} = (\Delta\rho/\rho)_{max} + (\Delta\rho/\rho)_{TLmin} + (\Delta\rho/\rho)_{Tmin}$$

$$r_{TL} = (\Delta\rho/\rho)_{TLmin}/(\Delta\rho/\rho)_{max},$$

$$r_T = (\Delta\rho/\rho)_{Tmin}/(\Delta\rho/\rho)_{max}.$$

These two anisotropy ratios r_{TL} and r_T are the parameters showing the degree of preferred orientation. For the ideal planar, axial, point and random orientations, the following values have to be obtained:

ideal planar orientation:	$r_{TL}=0$ and $r_T=0$
ideal axial orientation:	$r_{TL}=0$ and $r_T=1$
ideal point orientation:	$r_{TL}=1$ and $r_T=1$
ideal random orientation:	$r_{TL}=1$ and $r_T=1$.

In axial orientation scheme, coaxial and radial alignments cannot be differentiated from these r_{TL} and r_T values. For point and random orientation schemes, both r_{TL} and r_T are equal to 1 and the two schemes cannot be differentiated.

Magnetoresistance $\Delta\rho/\rho$ can be employed as a parameter for characterizing the crystallographic structure of carbon materials [Hishiyama et al., 1991], as shown above. In order to discuss the structural changes with HTT on a carbon material, maximum transverse magnetoresistance $(\Delta\rho/\rho)_{max}$ can be employed as one of the parameters. When different carbon materials have to be compared in their structure, however, the value of crystallite magnetoresistance $(\Delta\rho/\rho)_{cr}$ has to be used. The correspondence of this $\Delta\rho/\rho$ to other structural parameters measured mostly by X-ray diffraction was studied recently [Iwashita et al., 1997], including other electronic properties, electrical resistivity, Hall coefficient, etc.

Electromagnetic properties, electrical resistivity ρ and Hall coefficient R_H, as well as magnetoresistance $\Delta\rho/\rho$, have been measured on various carbon materials in order to understand not only the energy band structure but also the crystallographic structure [Hishiyama et al., 1991]. A typical example of the dependences of these characteristics is shown on a needle-like coke in Fig. 1-19.

(c) Transmission electron microscopy

For the study of the structure of carbon materials, different techniques of transmission electron microscopy (TEM) have been employed: bright-field image, selected area electron diffraction, dark-field images using different diffraction lines and lattice fringes [Oberlin, 1989]. In Fig. 1-20, as an example, some images obtained by TEM techniques are shown on a flake which has been obtained by the graphitization under pressure.

The bright-field image (Fig. 1-20a) is the image by electron beam transmitted through specimen and gives an appearance of the specimen particle. At the place shown by "1" in the figure, a Moiré pattern is observed, which is due to a slight rotation of crystal planes and the existence of which indicates the formation of well-developed crystallographic planes, graphitic layers in the case of carbon materials. At the place "2", another contrast

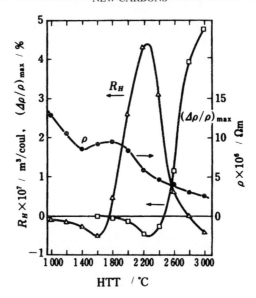

Figure 1-19   Dependences of electrical resistivity ρ, Hall coefficient R_H and maximum magnetore-sistance $(\Delta\rho/\rho)_{max}$ measured at liquid nitrogen temperature under a magnetic field strength of 1 T on HTT for a needle-like coke [Hishiyama et al., 1984].

is observed, which is called diffraction contrast and is due to a strong diffraction of the incident electron beam, here due to a strong diffraction by graphitic layers which are perpendicular to the picture.

The electron diffraction occurs as in the case of X-ray, but in TEM only the diffraction lines from the crystal planes almost parallel to the incident beam are observed because of the very short wavelength of the electron beam. Figure 1-20b shows the electron diffraction pattern observed on the particle in Fig. 1-20a. In this pattern, two sets of diffraction spots with hexagonal symmetry assigned to *100* and *110* diffraction are observed, indicating that graphite layers in most parts of this particle are perpendicular to incident beam. In addition, a pair of *002* diffraction spots is observed, which is due to the part "2" in the bright-field image where the graphite layers are parallel to the incident beam. In electron microscopy, the area giving diffraction can be limited and selected. When the diffraction pattern was obtained by selecting the area around the part "1", a quite similar pattern was obtained, except for a pair of *002* spots.

The lattice images can be obtained by using transmitted and one of the diffracted beams. For carbon materials, the *002* diffraction line was mostly used (*002* lattice fringes). Figure 1-20c is the *002* lattice fringes observed at the place "2" in Fig. 1-20a, where the layers are extended in parallel over a long range.

The image of the particle can be obtained by using only one diffraction line (dark-field image). When one of the diffracted beams appeared in the diffraction pattern (Fig. 1-20b) the places that gave the diffracted beams were lightened: using 002 spot the place "2" was lightened (*002* dark-field image) and using either *100* or *110* spots most of this particle was lightened (*100* or *110* dark-field image, respectively).

This dark-field image observation was successfully applied to the texture analysis in carbon spheres which were synthesized from the mixture of polyethylene and polyvinylchloride under pressure [Hishiyama et al., 1982]. Figure 1-21 shows a set of *002* dark-field images obtained at different positions of the aperture on a cross-section of a number of spheres which are mounted in a resin. By changing the position of the aperture the part lightened in a sphere changes gradually. From detailed analysis of these

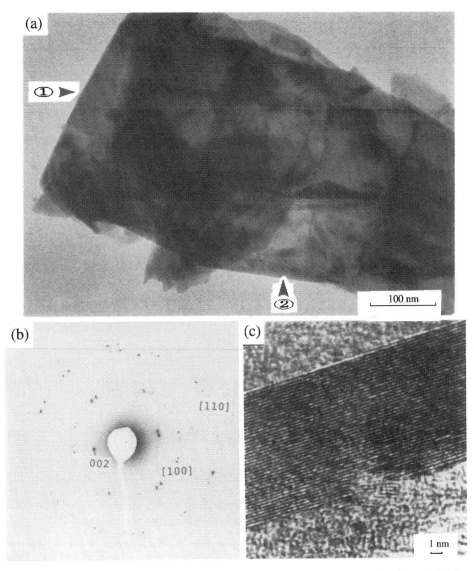

Figure 1-20 Transmission electron micrographs on a flake of well-crystallized carbon: (a) bright-field image; (b) electron diffraction pattern; (c) *002* lattice fringe image. [Courtesy of Mme A. Oberlin.]

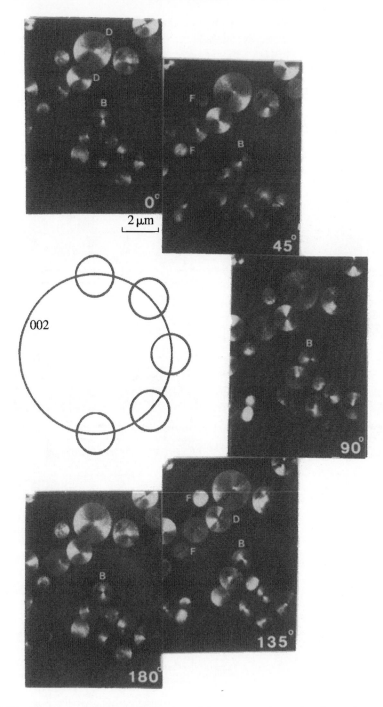

Figure 1-21   Change in *002* dark-field images of the cross-sections of carbon spheres with the rotation of aperture along *002* diffraction ring [Hishiyama et al., 1982].

observations, the radial texture of hexagonal layers was concluded. An example of a *110* dark-field image of well-graphitized film derived from a polyimide by high-temperature treatment [Bourgerette et al., 1993] is shown in Fig. 1-22, which reveals well-developed graphitic layer planes perpendicular to electron beam with Moiré patterns.

(d) Scanning electron microscopy

Scanning electron microscopy (SEM) has been commonly used to observe the morphology and surface of the materials, including carbon materials. In order to obtain correct information on carbon materials, the following points have to be taken into account.

In materials that consist of an element with a low atomic number, such as carbon, the electron beam can penetrate and so the SEM images obtained are the accumulation of secondary electrons reflected from all the parts along the depth of penetration. As a consequence, the image is not obtained from the exact surface of the carbon material. Therefore, using a low acceleration voltage for the electron beam is recommended [Yoshida and Hishiyama, 1987]. Scanning electron micrographs taken using different acceleration voltages from 40 to 1 kV on a cross-section of a single filament of 1500°C-treated mesophase-pitch-based carbon fiber are compared in Fig. 1-23. Differences in the appearance of the micrographs are obvious. From these studies, an acceleration voltage of

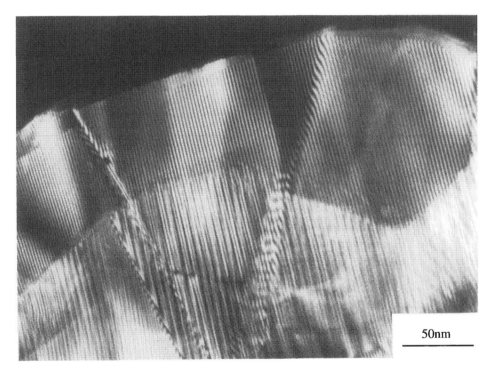

50nm

Figure 1-22 *110* dark-field image of a well-graphitized carbon film derived from a polyimide films. [Courtesy of Mme A. Oberlin and Mr C. Bourgerette.]

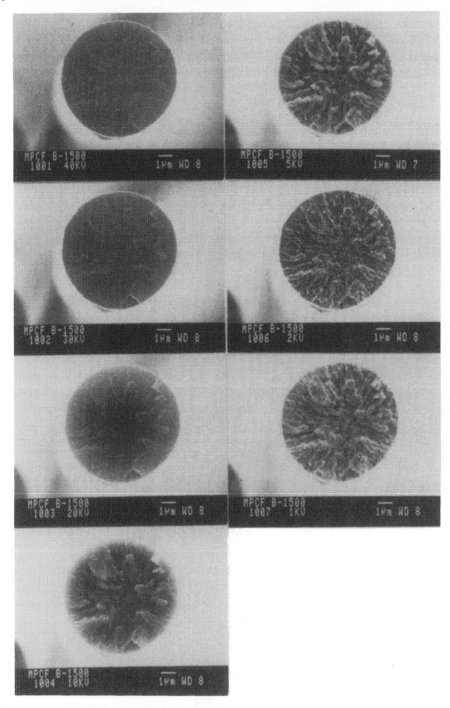

Figure 1-23 Scanning electron micrographs taken using different acceleration voltage on the fractured cross-section of a mesophase-pitch-based carbon fiber. [Courtesy of Prof. Y. Hishiyama and Dr A. Yoshida of Musashi Institute of Technology.]

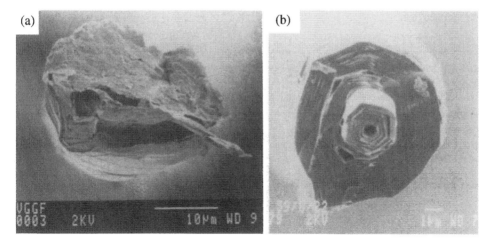

Figure 1-24 Scanning electron micrographs of cross-sections of vapor-grown carbon fibers, prepared using a knife at room temperature (a) and by applying a shock at liquid nitrogen temperature (b). [Courtesy of Prof. Y. Hishiyama and Dr A. Yoshida of Musashi Institute of Technology.]

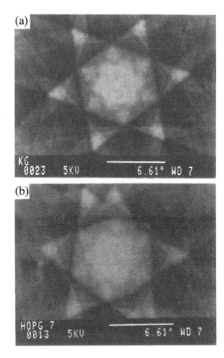

Figure 1-25 Electron channeling contrast image of a flake of kish graphite (a) and of highly oriented pyrolytic graphite (HOPG) (b). [Courtesy of Prof. Y. Hishiyama and Dr A. Yoshida of Musashi Institute of Technology.]

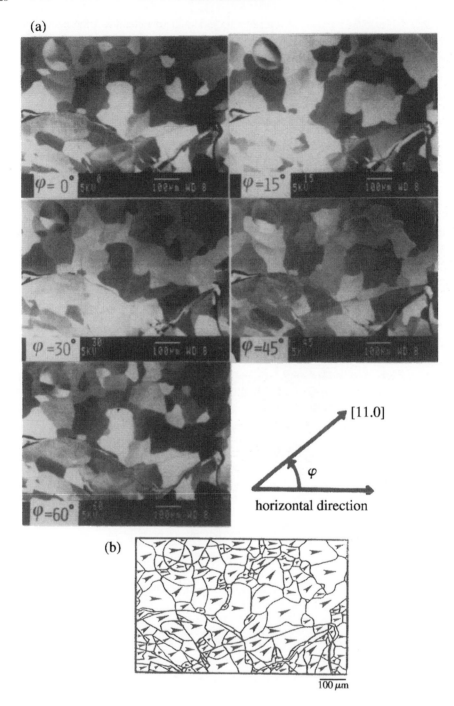

Figure 1-26 Electron channeling contrast images of highly oriented pyrolytic graphite (HOPG) as a function of rotation angle φ (a) and a-axis orientation of each single crystal domain in the flake (b). [Courtesy of Prof. Y. Hishiyama and Dr A. Yoshida of Musashi Institute of Technology.]

about 2 kV, less than 5 kV, was found to be essential to understand the exact surface texture of carbon materials.

It has been pointed out that special care has to be taken to prepare the fractured surface for SEM observation [Yoshida and Hishiyama, 1989]. This is particularly important for well-crystallized carbon materials, such as highly oriented graphites (Chapter 2) and vapor-grown carbon fibers (Chapter 5), because of easy cleavage. The fracture of sample carbon at liquid nitrogen temperature by giving a shock through frozen acetone was strongly advised. In Fig. 1-24, scanning electron micrographs of a cross-section of well-graphitized vapor-grown carbon fiber are compared; micrograph (a) is on the cross-section fractured by a knife at room temperature and (b) on that prepared at liquid nitrogen temperature. The latter represents the characteristic annual-ring texture and also the polygonization of vapor-grown carbon fibers (Chapter 5).

Information on the alignment of graphitic layers in a plane, particularly in flakes of highly oriented graphite, can be obtained by electron channeling techniques in a scanning electron microscope [Yoshida and Hishiyama, 1992]. This technique uses the fact that the backscattering efficiency of the primary electron beam changes with the angle between the incident beam and the crystal lattice of the specimen, and decreases discontinuously at the angle satisfied by the Bragg condition [Joy et al., 1982]. This discontinuity in backscattering efficiency for each crystal plane is observed as an electron channeling pattern, as shown on a flake of kish graphite (Chapter 2) in Fig. 1-25a. The pattern with well-defined hexagonal symmetry reveals highly crystallized graphite layers. From the sample with a little less crystallinity, a distorted pattern is observed (Fig. 1-25b), as will be discussed in detail in Chapter 2.

Both the Bragg angle for the crystal plane and the backscattering efficiency depend on the acceleration voltage of the primary electron beam. If an acceleration voltage of 5 kV is selected and the sample graphite is declined by $7.5°$ from the horizontal, the crystallites are brightened, of which the crystallographic direction of [11.0] coincides with the declining direction of the sample flake. Therefore, if the sample flake is rotated by keeping this declining angle and electron channeling contrast images are observed as a function of rotation angle φ, the size and shape of crystallites in the sample flake can be measured. The series of images in Fig. 1-26a shows an example of electron channeling contrast on a flake of HOPG (Chapter 2), the image being reproduced by every $60°$ rotation. By electron channeling contrast the orientation of small crystallites can be visualized, as shown in Fig. 1-26b, where a region with a homogeneous contrast corresponds to a crystallite and the regions with different contrast have different orientation of the a-axis of the graphite layer.

CHAPTER 2

Highly Oriented Graphites

2.1. What are Highly Oriented Graphites?

Basic structural units, from which most carbon materials used in industry and everyday life are composed, are the stacked layers of carbon hexagons [Oberlin, 1984, 1989; Oberlin et al., 1998]. An extreme case, i.e. with very large-sized layers stacked in large numbers with regularity, is a single crystal of graphite. In practice, however, single crystals with large sizes are difficult to obtain, and it is almost impossible to obtain those with a layer size of more than a few square millimeters. Graphite has a typical layered structure and, in consequence, very strong anisotropy in its properties, e.g. very easy cleavage along the layer, and an electrical conductivity that is high along the layer but very poor perpendicular to the layer. Therefore, it is impossible to obtain single crystals with a certain thickness.

There are only two ways of finding single crystals of graphite, in natural graphite ores and kish graphite. The resources of high-quality natural graphite are limited to Sri Lanka, Madagascar and China. In these natural ores, only occasionally are single crystals of a certain size found. The formation of kish graphite is artificial; it is formed by the precipitation of supersaturated carbon from molten iron. Kish graphite thus precipitated cannot be of a large size, but some samples have been found to have very high crystallinity, which can be said of single crystals.

The alternatives for single crystal graphite are the synthetic graphites of which layers are highly oriented, such as highly oriented pyrolytic graphite (HOPG) and graphite films derived from some organic polymer precursors, such as polyimides. These two materials cannot be single crystals. They have an extremely high degree of orientation of the c-axis of small graphite crystals (crystallites) along one direction, usually the normal of the flakes and films, but the a-axes of graphite crystallites in the flakes and films are random.

In the present chapter, three kinds of highly oriented graphites, kish graphite, highly oriented pyrolytic graphite and graphite films derived from polyimides, are explained regarding their preparation, with particular emphasis on the control of a high degree of orientation, crystallographic structures and different properties.

2.2. Kish Graphite

(a) Preparation

The amount of carbon atoms that can be dissolved into metal iron depends strongly on the temperature. Some of the carbon dissolved into molten iron at high temperatures is incorporated into the crystal lattice of iron and forms alloys, thus giving different steels. However, another part of the carbon precipitates as a separated phase from iron, in most cases as graphites. These graphites precipitated from molten iron at high temperatures are called kish graphite. In the course of the production of iron and alloys, a rather large amount of kish graphite is obtained as a byproduct, but this does not usually have a very high crystallinity, because it depends strongly on the precipitation conditions, as in normal crystallization. These kish graphite flakes initially attracted the attention of scientists, because some of the flakes were found to have single crystal nature when they were produced at a very high temperature, as high as the temperature at which iron evaporates.

In these kish graphite flakes formed either during the cooling of molten iron or by evaporation of iron, a relatively large amount of iron remains as an impurity. In order to obtain highly crystallized graphite flakes which can be regarded as single crystals, therefore, purification in a flow of halogen gas at high temperatures is essential.

Some of the flaky particles of kish graphite are shown in Fig. 2-1. As can be seen, the particles of kish graphite have an irregular shape and are very thin.

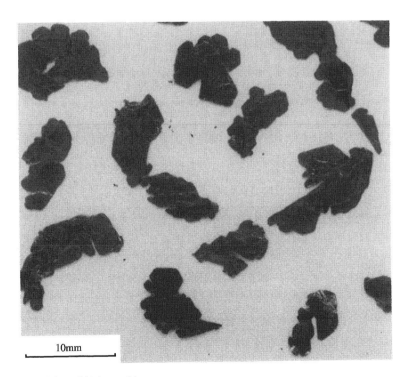

10mm

Figure 2-1 Flakes of kish graphite.

(b) Structure and properties

The high crystallinity of the particles of kish graphite prepared through the evaporation of iron at high temperatures and then purified in the flow of a halogen gas at high temperatures was confirmed from observations by high-fidelity scanning electron microscopy on the edge surface of the flake and electron channeling pattern on the surface [Yoshida and Hishiyama, 1987, 1992]. Their edge surfaces are constructed from regular stacking of layers, as shown in Fig. 2-2a and b by SEM using a low acceleration voltage of 3 kV. The surfaces along the layers give a well-organized electron channeling pattern, as shown in Fig. 2-3, indicating that axes of graphite crystallites are aligned in one direction, i.e. a single crystal, which shows a sharp contrast to other highly oriented graphites, HOPG and graphite films, as will be explained in the following sections. In the electron channeling contrast image, no contrast was observed and its brightness changes in every 60° rotation of the flake.

On examining the crystallinity of a large number of flakes of kish graphite by the measurement of electromagnetic properties, however, each flake was found to have a wide range of perfection of the crystal. Table 2-1 shows the resistivity ratio $\rho_{300K}/\rho_{4.2K}$, i.e. the

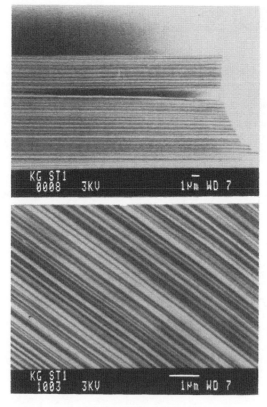

Figure 2-2 Fractured edge surface of kish graphite. [Courtesy of Prof. Y. Hishiyama and Dr A. Yoshida of Musashi Institute of Technology.]

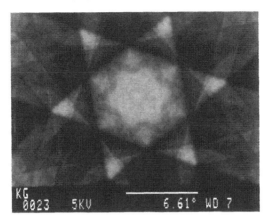

Figure 2-3   Electron channeling pattern of flake surface of kish graphite. [Courtesy of Prof. Y. Hishiyama and Dr A. Yoshida of Musashi Institute of Technology.]

ratio of electrical resistivity at room temperature (300 K) to that at liquid helium temperature (4.2 K), which has been used as a parameter to measure the structural

Table 2-1. Resistivity ratio $\rho_{300K}/\rho_{4.2K}$, maximum magnetoresistance $(\Delta\rho/\rho)_{max}$ at liquid nitrogen temperature under 1 T and anisotropy ratios r_{TL} and r_T for different flakes of kish graphite [Inagaki and Hishiyama, 1994]

Sample	$\rho_{300K}/\rho_{4.2K}$	$(\Delta\rho/\rho)_{max}$ (%)	r_{TL}	r_T
KG12	47.6	3880	0.0037	0.0054
KG13	35.7	3680	0.0069	0.0099
KG3	35.6	3460	0.0071	0.0091
KG18	34.5	3510	0.0033	0.0071
KG100	30.2	3450	0.0068	0.0087
KG836	29.6	3250	0.0064	0.0107
KG6	28.8	2900	0.0070	0.0095
KG31	26.8	2840	0.0062	0.0088
KG30	20.9	3000	0.0038	0.0084
KG15	20.0	2410	0.0096	0.0196
KG17	19.2	2580	0.0073	0.0099
KG1	19.1	2550	0.0081	0.0118
KG692	18.9	2550	0.0036	0.0200
KG127	12.7	2200	0.0171	0.0111
KG4	11.9	2000	0.0074	0.0093
KG40	9.01	1860	0.0105	0.0138
KG5	8.86	1900	0.0049	0.0174
KG2	7.96	1380	0.0042	0.0081
KG11	7.63	1410	0.0072	0.0124
KG9	7.52	1280	0.0084	0.0248
KG20	6.62	1260	0.0121	0.0141
KG60	6.21	1120	0.0107	0.0189
KG19	5.95	1000	0.0151	0.0261
KG8	4.81	1250	0.0145	0.0196
KG835	4.71	781	0.0050	0.0071

perfection of graphite, and maximum transverse magnetoresistance $(\Delta\rho/\rho)_{max}$ measured at liquid nitrogen temperature (77 K) under a magnetic field perpendicular to the flakes with a strength of 1 T, which has been used as a measure of the development of three-dimensional graphitic structures [Hishiyama et al., 1991; Inagaki and Hishiyama, 1994]. The larger values of $\rho_{300K}/\rho_{4.2K}$ and $(\Delta\rho/\rho)_{max}$ reveal a more perfect crystal structure.

The flakes of kish graphite have a wide range of crystallinity, from 50 to 4 in $\rho_{300K}/\rho_{4.2K}$, even though they have been prepared through high-temperature evaporation of iron (Table 2-1). Most of these flakes show very small values of r_{TL} and r_{P}, indicating a high degree of planar orientation. In kish graphite flakes formed through precipitation from molten iron, the range of structural perfection of each flake is expected to be extended towards much lower values of $\rho_{300K}/\rho_{4.2K}$ and $(\Delta\rho/\rho)_{max}$.

On these kish graphites with high crystallinity, Shubnikov-de Haas oscillation in magnetoresistance, which is characteristic for single crystals, was observed at temperatures as low as 4.2 K [Hishiyama and Kaburagi, 1987]. The dependences of $\Delta\rho/\rho$ on magnetic field strength B for some kish graphite flakes with different values of $\rho_{300K}/\rho_{4.2K}$ are shown in Fig. 2-4. The flake with the higher resistivity ratio shows more marked Shubnikov-de Haas oscillation.

Absolute thermoelectric power (TEP) is also very sensitive to the crystal perfection. In Fig. 2-5, the changes in TEP with temperature are compared with different flakes of kish graphite and HOPG with relatively high values of $\rho_{300K}/\rho_{4.2K}$, which are indicated for each sample [Hishiyama and Ono, 1985; Sugihara et al., 1986]. On all samples shown, three anomalies in TEP due to a phonon-drug, i.e. a sharp minimum around 35 K, a small but sharp peak around 10 K and a shallow minimum around 3 K, are found to overlap with the dependence of TEP on temperature due to scattering. The phonon-drug phenomena in graphite crystals are discussed in relation to the crystallinity. For example, the minimum

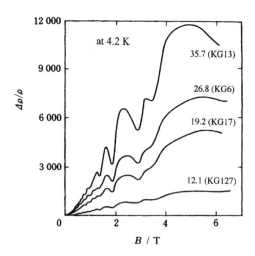

Figure 2-4 Dependences of magnetoresistance $\Delta\rho/\rho$ at 4.2 K on magnetic field strength B for kish graphite flakes with different resistivity ratios $\rho_{300K}/\rho_{4.2K}$ [Hishiyama, 1987]. The sample codes shown in parentheses refer to data in Table 2-1.

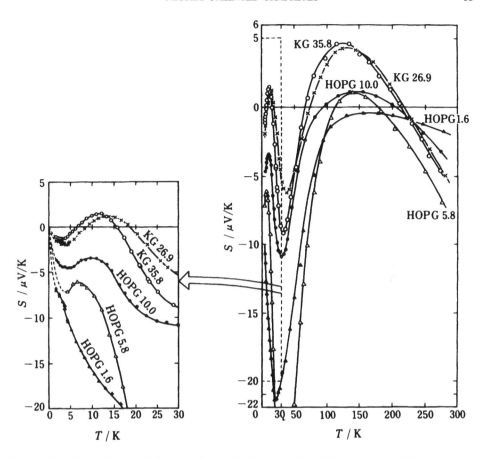

Figure 2-5 Dependences of absolute thermoelectric power S on the temperature of measurement for different kish graphite flakes and HOPG samples with different values of the resistivity ratio $\rho_{300K}/\rho_{4.2K}$ [Hishiyama and Ono, 1985].

at 35 K, which is known to be due to the interaction between the phonons of lattice vibration within the graphite layer and carriers, depends strongly on the crystallinity, becoming deep with the decrease in $\rho_{300K}/\rho_{4.2K}$, i.e. the decrease in crystal perfection.

2.3. Highly Oriented Pyrolytic Graphite (HOPG)

(a) Preparation

Hydrocarbon gases, such as methane and propane, give a deposition of carbon at high temperatures on a substrate, which has been called pyrolytic carbon and used for carbon coating of nuclear fuel particles in a reactor. It has been extensively studied to control the structure and texture of pyrolytic carbons formed by changing the deposition conditions, such as the kind of hydrocarbon gas, its concentration and flow rate, temperature of deposition and the geometry of the furnace [Bokros, 1969]. When the graphite substrate was used as a heater by passing electric current directly through it, pyrolytic carbon deposited close to the substrate, in other words, deposited in the beginning of the process, was found to be graphitized more than the part near the surface [Blackman and Ubbelohde, 1962; Noda et al., 1962; Inagaki, 1963a]. This inside part was thought to be formed by rather large compressive stress in addition to the higher temperature than that outside, because of a large temperature gradient across the thickness of the pyrolytic carbon deposited. This may provide a method by which to produce HOPG [Moor, 1973, 1982].

Figure 2-6 shows a typical procedure used to produce HOPG, which consists of annealing at high temperatures such as 3400–3600°C under compression. In order to attain high crystallinity and a high degree of preferred orientation of the graphite layers, a certain structure and texture is required for the starting pyrolytic carbons with a growth cone texture, two typical textures being known as regenerative and singularly nucleated [Bokros, 1969]. The first step in the preparation shown in Fig. 2-6 is hot-pressing at 2800–3000°C, by which the growth cone texture is destroyed and the preferred orientation improved, as evaluated by mosaic spread (MS). A similar MS value was found to be obtained by either elongation or twisting of pyrolytic carbons. After this hot-pressing, the samples are easily cleaved and their surfaces give mirror reflections, but their physical properties are usually far from those observed on a single crystal of graphite. The second step of the preparation, annealing under pressure at high temperatures above 3400°C, is essential to improve the physical properties. At such high temperatures, graphite has some

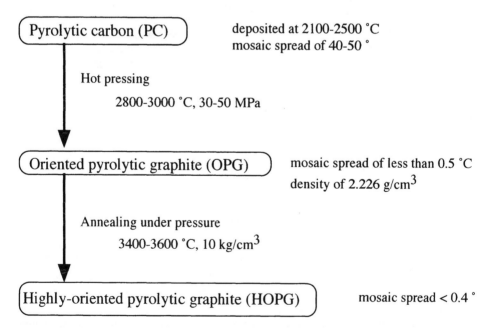

Figure 2-6   Procedure for the production of HOPG.

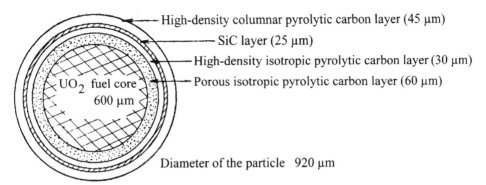

High-density columnar pyrolytic carbon layer (45 μm)

SiC layer (25 μm)

High-density isotropic pyrolytic carbon layer (30 μm)

Porous isotropic pyrolytic carbon layer (60 μm)

UO$_2$ fuel core
600 μm

Diameter of the particle 920 μm

Figure 2-7 Scheme of structure in nuclear fuel particles coated by carbon layers with different textures.

ductility [Green, 1953; Narisawa et al., 1994; Souma et al., 1995] and so the compressive stress is not necessarily very high.

The same deposition process was applied to the fine particles flowing in a gas stream, in order to produce the carbon-coated nuclear fuel particles of uranium compounds [Bokros, 1969]. An example of the construction of the cross-section of the particle is shown in Fig. 2-7, where the carbon layers with well-oriented and random textures are formed deliberately. These different textures of carbon layers are controlled by changing the deposition conditions. It was also successfully employed in the carbon coating of substrates with complicated shapes, a noted example being a heart valve [Bokros et al., 1972].

If very minute iron particles with a size less than 10 nm are on the substrate during the pyrolysis of hydrocarbon gases, fibrous carbon is formed, vapor-grown carbon fibers, which will be described in Chapter 4. By employing a high concentration of hydrocarbon gas, carbon blacks are produced which have been used in rubbers for reinforcement and in black inks for printing, described as one of the classical carbon materials in Chapter 1.

(b) Structure and properties

As described above, the structure of pyrolytic carbons and, in consequence, their properties depend primarily on the preparation condition. The resistivity ratio $\rho_{300K}/\rho_{4.2K}$, maximum transverse magnetoresistance $(\Delta\rho/\rho)_{max}$ and anisotropy ratios r_{TL} and r_T of HOPG prepared under compression at the temperature shown in the sample code are listed in Table 2-2 [Inagaki and Hishiyama, 1994]. In this table, the data on pyrolytic graphites prepared under elongation and twisting (shown by E and T, respectively) are also shown.

The crystallinity of HOPG depends strongly on the preparation conditions, particularly the temperature of the final heat treatment. However, it has to be pointed out here that HOPG heated up to 3600°C, near the evaporation temperature of graphite, shows that the crystallinity measured by $\rho_{300K}/\rho_{4.2K}$ and $(\Delta\rho/\rho)_{max}$ cannot overcome those of kish graphite shown in Table 2-1. Twisting and elongation at a high temperature are an effective way, along with compression, to improve the orientation of crystallites and so the crystallinity of samples, but the temperature is primarily important.

Table 2-2. Resistivity ratio $\rho_{300K}/\rho_{4.2K}$, maximum magnetoresistance $(\Delta\rho/\rho)_{max}$ at liquid nitrogen temperature under 1 T and anisotropy ratios r_{TL} and r_T for HOPG [Inagaki and Hishiyama, 1994]

Sample	$\rho_{300K}/\rho_{4.2K}$	$(\Delta\rho/\rho)_{max}$ (%)	r_{TL}	r_T
HOPG 3600-1	4.50	1210	0.0160	0.0134
HOPG 3600-2	4.03	1110	0.0048	0.0058
HOPG 3600-3	2.83	812	0.0080	0.0092
HOPG 3200-2	1.60	356	0.0127	0.0154
HOPG 3300	1.35	416	0.0094	0.0077
HOPG 3100T-1	1.17	284	0.0067	0.0120
HOPG 3200-1	1.14	304	0.0194	0.0187
HOPG 3100T-2	1.13	296	0.0111	0.0142
HOPG 3100E	1.06	254	0.0169	0.0283
HOPG 2760E-1	0.798	118	0.0033	0.0398
HOPG 2760E-2	0.709	101	0.0038	0.0416
HOPG 2800T-1	0.546	16.2	0.0235	0.0435
HOPG 2800T-2	0.502	8.92	0.0289	0.0477

T: under twisting; E: under elongation.

In Fig. 2-8, a representative electron channeling pattern and its contrast image are shown on a HOPG [Yoshida and Hishiyama, 1992]. This channeling pattern is a little

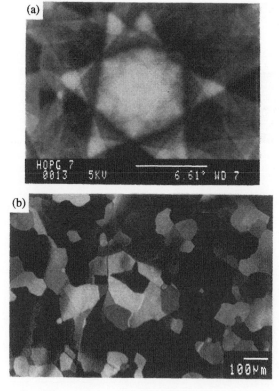

Figure 2-8 Electron channeling pattern (a) and channeling contrast (b) of a flake of HOPG. [Courtesy of Prof. Y. Hishiyama and Dr A. Yoshida of Musashi Institute of Technology.]

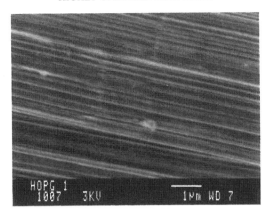

Figure 2-9 Scanning electron micrograph of fractured surface of a flake of HOPG. [Courtesy of Prof. Y. Hishiyama and Dr A. Yoshida of Musashi Institute of Technology.]

distorted, in contrast to that observed on the flake of kish graphite shown in Fig. 2-3, indicating the aggregation of crystallites with slightly different orientations. The detailed analysis of an electron channeling contrast image with different orientations shows clearly that HOPG is composed from crystalline domains ("crystallites"); in a domain a c-axis is unique, i.e. single crystal nature, but their a-axes have different orientations with each other. The SEM on the cross-section, which is obtained by breaking the HOPG block at liquid nitrogen temperature by applying a shock, shows a regular stacking of the basal planes of graphite, as shown in Fig. 2-9 [Yoshida and Hishiyama, 1987]. A few streaks perpendicular to the stacked layers are observed, as reasonably expected to come from the growth cone texture in the starting pyrolytic carbons and which might cause the distortion in orientation.

As explained above, HOPG is inferior in crystallinity to kish graphite, but has an advantage in being able to have a large size. This is the reason why HOPG has been used as the samples for various fundamental scientific researches, instead of single crystals. It developed a large application as a monochromator for X-ray and neutron diffractometry [Moor, 1982]. The primary demand on a monochromator of radiation is a high degree of orientation of well-crystallized graphite layers, the mosaic spread of which, in practice, must be less than 0.2°.

2.4. Graphite Film Derived from Polyimides

(a) Preparation

Polyimides have been developed as thermoresistant polymers, and have been used in different fields, especially in the field of electronics. Some polyimide films were found to give a graphite film with rather high crystallinity through a simple heat treatment under atmospheric pressure [Inagaki et al., 1998b]. Because of the practical and promising applications of polyimide films, there have been commercially available films with different molecular structures, giving a wide variety of structure in the carbon films after heat treatment at high temperatures, from highly crystalline graphite to amorphous glass-like carbon films. This is a typical case where the molecular structure of organic precursors and the texture of their polymer films govern the structure and texture of the resultant carbon films, i.e. the crystallinity and preferred orientation of hexagonal layers.

Aromatic polyimides (D in Fig. 2-10) are formed by a reaction of tetracarboxylic dianhydride (A) with equimolar aromatic diamine (B) in either N-methyl-2-pyrrolidone (NMP) or N,N'-dimethylacetamide (DMAc), through dehydration of the intermediate product of polyamic acid (C) by either thermal or chemical methods (imidization), as shown in Fig. 2-10.

It has been shown in studies on carbonization and graphitization of these polyimide films that their molecular structure is considered to consist of two constituent parts, imide and bridging parts (D). From Fig. 2-10, it is clear that the former comes from the anhydride and the latter from the diamine used. The polyimides, therefore, can be characterized by coupling the names of starting anhydride and diamine, such as PMDA/ODA, which is the polyimide made from pyromellitic dianhydride (PMDA) and 4,4'-oxydianiline (ODA).

In Fig. 2-11, the imide molecules which have been studied on carbonization and graphitization are summarized by aligning the imide part in the perpendicular direction and the bridging part in the horizontal one. There are various commercially available polyimide films and so the trade names are also cited on the corresponding structures of the main constituent imide molecules. It was experimentally shown that, for instance, the commercially available polyimide film "Kapton" is constituted from the imide molecules of PMDA/ODA, but has a quite different behavior during carbonization and graphitization from the film of the same molecules of PMDA/ODA prepared in the laboratory. Therefore, trade names such as Kapton were used when the commercial films were employed as samples and the structural names such as PMDA/ODA when the laboratory-made films were used.

For a Kapton film with 25 μm thickness, the changes in film weight and shrinkage along the film with carbonization temperature are shown in Fig. 2-12 and the changes in the composition of decomposition gas during carbonization are shown in Fig. 2-13 [Inagaki et al., 1989, 1992a].

The results in Figs 2-12 and 2-13 show that the carbonization of the polyimide Kapton proceeds in two steps, the first step in the rather narrow temperature range of 550–650°C, showing an abrupt weight decrease associated with the evolution of a large amount of carbon monoxide and pronounced shrinkage along the film, and the second step with small weight loss, evolution of small amounts of methane, hydrogen and nitrogen, and little shrinkage over a temperature range from 700°C to more than 1000°C. The first step of decomposition is seen to be due mainly to a breakage at carbonyl groups in the imide part. The etheric oxygen was supposed to be released at the end of the first step, from the comparison of the result on the molecule without etheric oxygen (e.g. PMDA/PPD in Fig. 2-11). The release of nitrogen in the second step of carbonization was found to continue up to high temperatures above 2000°C, leaving large amount of pores in the film if it was heated continuously up to 2400°C.

The structural change in the second step of carbonization was found to reflect the electrical properties of the film [Inagaki et al., 1992a]. In Fig. 2-14, the change in electrical conductivity along the film at room temperature with carbonization temperature is plotted on different laboratory-made polyimide films. All films, including PMDA/ODA (corresponds to Kapton), show a pronounced increase in conductivity, more than one order of magnitude, between 700 and 800°C, although only a small weight loss and shrinkage

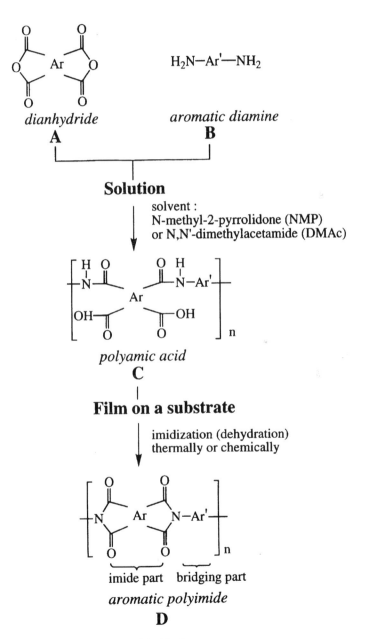

Figure 2-10 Preparation process of polyimide films.

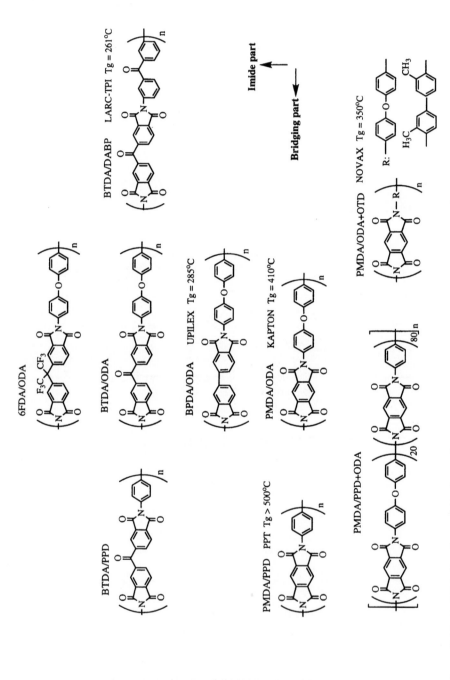

Figure 2-11 Imide molecules used for the studies on carbonization and graphitization [Inagaki et al., 1998b].

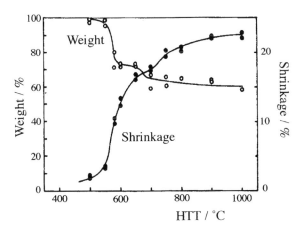

Figure 2-12 Changes in weight and shrinkage of polyimide Kapton films with a thickness of 25 μm with HTT [Inagaki et al., 1989].

are observed (refer to Fig. 2-12). Further carbonization up to 1100°C gives another increase in conductivity, roughly one more order of magnitude. The higher conductivity of PMDA/PPD film (see Fig. 2-11) than other films is explained by the higher orientation of the constituent imide molecules with better planarity. The slightly lower conductivity of 6FDA/ODA film is also expected from the steric structure of its precursor molecule (see Fig. 2-11).

The facts that methane and hydrogen released and electrical conductivity increased in the second step suggest that the carbonization process occurs mainly in the second step,

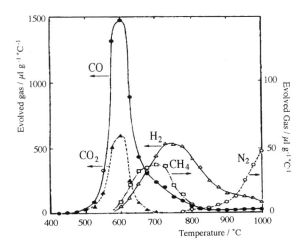

Figure 2-13 Spectra of decomposition gases from polyimide Kapton film [Inagaki et al., 1992a].

and the first step may be pyrolysis of polyimide molecules. It is worthwhile mentioning that homogeneous and dense carbon films are obtained even though weight loss is as large as 40% and linear shrinkage as large as 22%. No cracks are observed even under SEM.

In Fig. 2-15, some representative images of 002 lattice fringes are shown on a polyimide film, which gives well-crystallized graphite films at high HTT, heat-treated at different temperatures above 1500°C [Bourgerette et al., 1992, 1993, 1995]. Figure 2-16 is a schematic illustration of the nanometric texture change in their cross-section with HTT.

Up to 2200°C, it is clear that tiny pores are formed between the carbon layer stacks and they are flattened along the film surface. At 2300°C, these flattened pores collapse so that largely extended flat and perfect layers are suddenly observed. Above this temperature, a large grain texture is produced, accompanied by a sudden progress of graphitization, which is also shown by X-ray diffraction and galvanomagnetic property measurements, as will be shown later in Fig. 2-19. The stacking of carbon layers is turbostratic after heat treatment at 2200°C, but above 2300°C it is improved, being accompanied by pore collapsing, to a graphitic one.

From an initial thickness of 25 μm, the film becomes thinner with the increase in HTT, as shown in Fig. 2-17 [Bourgerette et al., 1992, 1995] on different polyimide films. For Kapton, the first abrupt decrease at 550°C corresponds to a massive release of oxygen as CO and CO_2 (Figs 2-12 and 2-13). However, the release of CH_4, H_2 and also N_2 in the

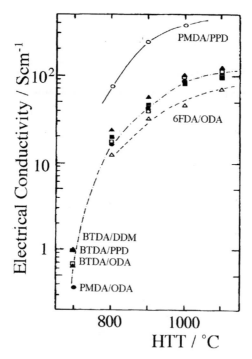

Figure 2-14 Changes in electrical conductivity along the film at room temperature with HTT [Inagaki et al., 1992a]. ○: PMDA/PPD; ▲: BTDA/PPD; ●: PMDA/ODA; ❑: BTDA/ODA; △: 6FDA/ODA; ■: BTDA/DDM.

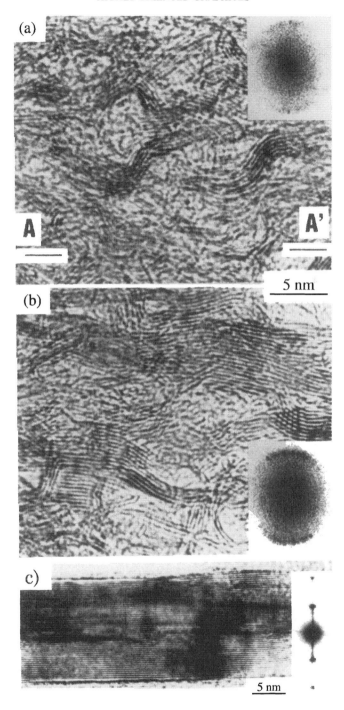

Figure 2-15 002 lattice fringes of carbon films prepared from a polyimide film, Kapton, at different temperatures: (a) 2250°C, (b) 2450°C and (c) 2550°C. [Courtesy of Mme A. Oberlin.]

HTT / °C

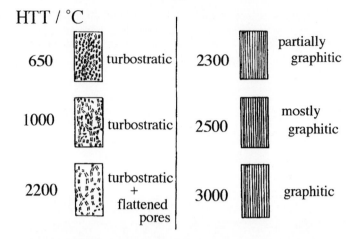

650	turbostratic	2300	partially graphitic
1000	turbostratic	2500	mostly graphitic
2200	turbostratic + flattened pores	3000	graphitic

Figure 2-16 Scheme of change in nanometric texture in the cross-section of carbon film derived from polyimide Kapton with HTT [Bourgerette et al., 1992].

second step of carbonization with a wide range of temperature from 700 to 1300°C does not lead to any change in thickness, and only gives a little shrinkage along the film (Fig. 2-12). Although no appreciable change in film thickness occurs, a considerable improvement in structure and texture occurs in the stacking order of the hexagonal carbon layers, as shown schematically in Fig. 2-16. Above 2000°C, particularly above 2300°C, the second thinning process is observed. This process corresponded to the large increase

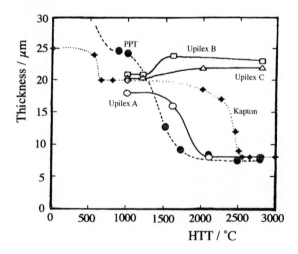

Figure 2-17 Changes in film thickness with HTT on carbon films derived from different polyimides [Bourgerette et al., 1995].

in optical anisotropy, the increasing contrast in the *002* lattice fringes of the flattened pores observed in TEM (Fig. 2-15). This corresponds to a sudden breakage of all defective areas near the edge of flattened pores, insuring complete collapsing of flattened pores and, as a consequence, lateral coalescence of the carbon layer stacks, which yields maximum compactness and a complete annealing of the layer distortions. The partial graphitization thus suddenly introduced does not give any further changes in thickness.

Other polyimide films, PPT and Novax, although the latter is not shown in Fig. 2-17, reach the same thickness as Kapton after high temperature treatment and achieved a high degree of graphitization, although their behaviors are slightly different from each other. On Upilex films, there are quite different changes in the thickness of carbonized films, film by film and also place by place in one film; some shrink as other polyimide films and reach a rather high graphitization degree (Upilex A in Fig. 2-17), and others show even expansion above 1500°C and give glass-like carbon films (Upilex B and C in Fig. 2-17).

In Fig. 2-18, the values of interlayer spacing d_{002} and crystallite sizes along the c-axis, $L_c(002)$ and $L_c(004)$ determined from both *002* and *004* diffraction lines, respectively, and also crystallite thickness along a-axis $L_a(110)$ determined from the *110* diffraction line are plotted against HTT on the carbon film prepared from Kapton film with 25 μm thickness by carbonization up to 900°C [Hishiyama et al., 1997]. The electromagnetic properties, electrical resistivity r, resistivity ratio r_{300K}/r_{77K}, Hall coefficient R_H and maximum transverse magnetoresistance $(\Delta\rho/\rho)_{max}$, are also shown on the same films in Fig. 2-19 [Hishiyama et al., 1997].

Only the *002* diffraction line was observed for the films heat-treated up to 1500°C. In the HTT range between 1700 and 2200°C, the d_{002} values measured from *002* and *004* lines are different, suggesting the presence of defects in the layers, and above 2300°C two values are the same. Above 2200°C, even the *006* line was observed clearly and above 2700°C both *004* and *006* lines showed splitting due to $K\alpha_1$ and $K\alpha_2$ radiation, which showed a marked improvement in crystallinity in the film, in accordance with marked increases in crystallite sizes in both L_c and L_a (Fig. 2-18b and c). Above 2200°C, the L_c values determined from *002* and *004* lines, $L_c(002)$ and $L_c(004)$, are different, suggesting the presence of some stacking disorder.

As shown in Fig. 2-19a, electrical resistivity ρ decreases rapidly at first and then gradually with increasing HTT, but in the HTT range between 2200 and 2300°C it drops discontinuously and then decreases again gradually. From the HTT dependence of ρ_{300K}/ρ_{77K} in Fig. 2-19b, three HTT ranges can be divided: 900–1200, 1300–2100 and above 2300°C. In the first range of HTT semiconductor-like conduction is observed, and in the second ρ_{300K}/ρ_{77K} is almost constant and weakly temperature dependent. Above 2300°C, the electrical conduction becomes semiconductive again, which agrees with the development of a graphitic structure. The ratio higher than 1.0 for the film heat-treated at 3000°C suggests good crystallinity. Its discontinuous change in the HTT range of 2100–2300°C is due to the abrupt transformation of the structure from turbostratic to graphitic.

The Hall coefficient R_H at 77 K and 1 T shows a maximum at the HTT of 2200°C, as shown in Fig. 2-19c. From the detailed measurements of its magnetic field dependence, R_H values on the films heat-treated below 2100°C were independent of magnetic field strength, being negative in the range of 1400–1600°C and positive at 1700–2100°C. Above

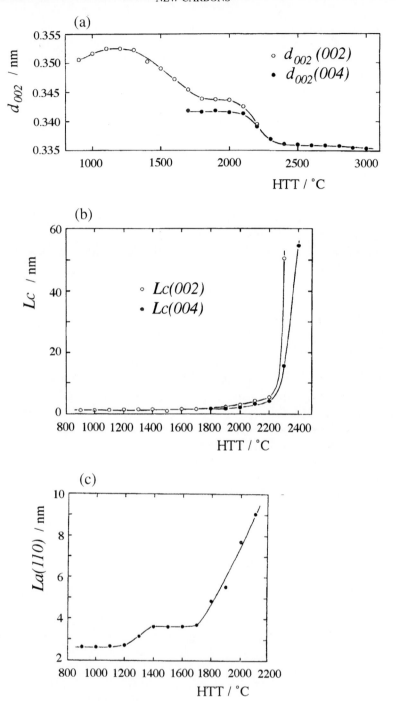

Figure 2-18 Changes in interlayer spacing d_{002} (a), crystallite sizes $L_c(002)$ and $L_c(004)$ determined from *002* and *004* lines, respectively, (b), and crystallite size along the *a*-axis $L_a(110)$ (c) of carbon film derived from Kapton with HTT [Hishiyama et al., 1997].

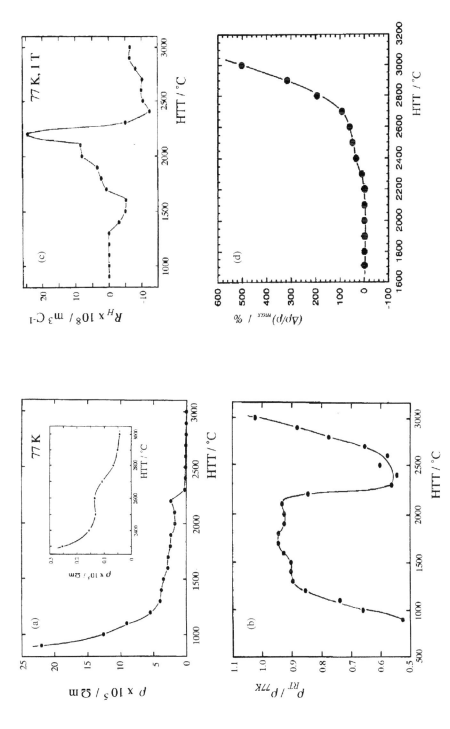

Figure 2-19 Changes in electrical resistivity ρ (a), resistivity ratio ρ_{RT}/ρ_{77K} (b), Hall coefficient R_H (c) and maximum magnetoresistance $(\Delta\rho/\rho)_{max}$ (d) of carbon film derived from Kapton with HTT [Hishiyama et al., 1997].

2300°C, R_H drops to a negative value and then gradually increases with the increase in HTT, its dependence on magnetic field being weak, but becoming similar to those for well-crystallized HOPG at higher HTT, suggesting the coexistence of two carriers, electrons and holes.

Maximum transverse magnetoresistance $(\Delta\rho/\rho)_{max}$ at 77 K and 1 T was not detected at HTT below 1500°C, negative up to 2200°C, then changed to positive and increased very quickly (Fig. 2-19d). The magnetic field dependences of $(\Delta\rho/\rho)_{max}$ as a function of HTT (Fig. 2-20) show a general trend common with graphitizing cokes [Hishiyama and Kaburagi, 1979]. Below a HTT of 2200°C, it was negative and increased its absolute value with the increase in magnetic field. At 2200°C, $(\Delta\rho/\rho)_{max}$ showed a trend of saturation at high field, suggesting the coexistence of a large amount of turbostratic structure with a small amount of graphitic structure. Above 2300°C it suddenly became positive and increased markedly under a magnetic field. It reached about 500% in the film heat-treated at 3000°C under a magnetic field of 1 T. The anisotropy ratios r_T and r_{TL} were as low as 0.05, showing a highly oriented texture in a planar orientation scheme of the films heat-treated at high temperatures. These changes in electromagnetic properties around 2200–2300°C correspond well to the abrupt changes in X-ray parameters shown in Fig. 2-18.

These abrupt changes in structural parameters, d_{002}, L_c, L_a and lattice fringes, and in electromagnetic properties, ρ, ρ_{300K}/ρ_{77K}, R_H and $(\Delta\rho/\rho)_{max}$, around 2200–2300°C were found to be very much related to the structural changes from turbostratic to graphitic, associated with the final departure of nitrogen atoms which were supposed to be substitutionally located in hexagonal carbon layers [Konno et al., 1997; Inagaki et al., 1998]. From the calculation using semi-empirical molecular orbital method, extra electron spins accompanied by nitrogen seemed to be localized around the C–N bond; in other words, nitrogen acting as an acceptor [Inagaki et al., 1998].

From studies on different polyimide films, three fundamental conditions for obtaining well-crystallized graphite films were concluded [Inagaki et al., 1991b]:

(1) flatness of the starting imide molecules,
(2) high degree of orientation of these flat molecules along the film,
(3) less disturbance in this orientation during carbonization and graphitization due to the out-gas of non-carbon atoms.

Factor (1) concerns the molecular structure of polyimide used as a precursor. As shown in Fig. 2-11, polyimides can have a wide variety of molecular structures, but not all produce high-quality graphite. From Kapton of which molecules were known to be flat [Isoda et al., 1981], for example, a graphite film with high crystallinity could be obtained by selecting the appropriate conditions of film preparation and heat treatment at high temperatures. The molecule of PMDA/PPD was completely flat and expected to give a highly graphitized film. In practice, however, the film made from PMDA/PPD was so brittle that it was broken easily only by a touching with the tip of a knife and, as a consequence, it had no application as plastic films. By using a small amount of tetramine with PMDA/PPD, a sufficiently flexible film was obtained, which was named PPT film

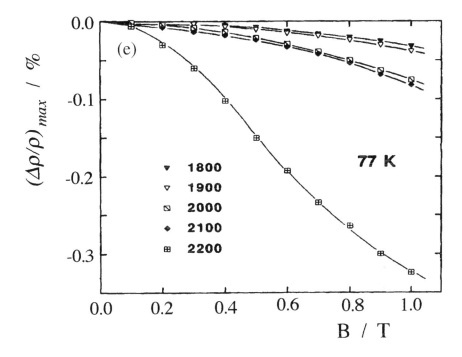

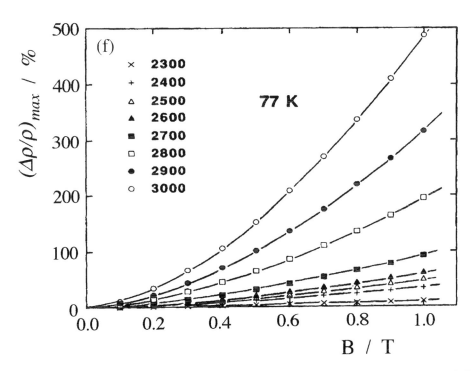

Figure 2-20 Dependences of maximum magnetoresistance $(\Delta\rho/\rho)_{max}$ on magnetic field strength B for the carbon films prepared from Kapton at different temperatures [Hishiyama et al., 1997].

[Nagata et al., 1991] and was confirmed to give highly graphitized films [Hishiyama et al., 1992, 1994; Kaburagi et al., 1995; Bourgerette et al., 1995]. Recently, a thin film of pure PMDA/PPD was found to be graphitized at temperatures as low as 2100°C [Hishiyama, 1998]. However, the film of Larc-TPI with a steric arrangement of constituent atoms in a molecule (Fig. 2-11) produced only a film of glass-like carbon even after heat treatment at temperatures as high as 3000°C [Hishiyama et al., 1993b].

Factor (2) can only be controlled by the conditions of film preparation, particularly by those at imidization and the selection of film thickness. Even starting with the molecules of PMDA/ODA (constituent molecule of Kapton), a high degree of graphitization in the film was not obtained unless a correct procedure for film preparation was employed [Inagaki et al., 1992c; Hatori et al., 1992c].

Factor (3) is related to the conditions during carbonization and graphitization. It was shown that the heating rate in the carbonization process had to be selected in the relation to the glass transition temperature T_g of the polyimide film used, because the orientation of molecules might be modified [Inagaki et al., 1994]. During graphitization, the parallel alignment of hexagonal carbon layers was supposed to be disturbed by the release of a small amount of nitrogen, for Kapton film it was recommended to keep at 2200°C before going to higher temperatures [Hishiyama, 1992].

(b) Quality of graphite films

The preparation conditions of the five films with rather high crystallinity, which are characterized by different techniques of electron channeling and galvanomagnetic property measurements, are summarized in Table 2-3 [Kaburagi et al., 1995]. The preparation conditions for each polyimide film were selected to give the highest crystallinity, taking into consideration the controlling factors mentioned above.

The size and a-axis orientation of the graphite layers aligned along the film surface were characterized using electron channeling contrast techniques. The electron channeling contrast images on the graphite films prepared from three polyimide precursors are shown in Fig. 2-21 [Hishiyama, 1992]. On all films selected, clear channeling contrast images are observed, including HOPG as shown in Fig. 2-8 and PG3200, revealing that these films

Table 2-3. Mosaic spread MS, resistivity ratio $\rho_{300K}/\rho_{4.2K}$, maximum magnetoresistance $(\Delta\rho/\rho)_{max}$ at liquid nitrogen and helium temperature, averaged anisotropy ratio r, mean free path λ and average grain size D determined from electron channeling contrast images for highly oriented graphite films [Kaburagi et al., 1995]

Sample	MS (°)	$\rho_{300K}/\rho_{4.2K}$	$(\Delta\rho/\rho)_{max}$ [77K, 1T]	$(\Delta\rho/\rho)_{max}$ [4.2K.1T]	r	λ (μm)	D (μm)
Kapton 3100	6.7	3.32	1254	7180	0.0113	3.5	8
Novax 3100	6.9	2.67	872	–	0.0173	–	8
PPT 3200	5.7	3.45	1206	5791	0.0170	2.6	8
PG 3200	8.6	1.60	338	1575	0.0186	1.5	5
HOPG 3600	0.9	5.50	1394	11252	0.0051	5.4	60

have a rather high crystallinity of graphite. It should be emphasized that the grains with a constant contrast correspond to the single crystal domain making its basal planes parallel to the film surface, i.e. parallel to the micrographs, and that if these grains are too small no contrast is observed in the micrograph. The boundaries between the grains with different contrasts are clearly defined more in the films derived from Kapton and PPT, being almost comparable to HOPG, than that from Novax, suggesting better crystallinity in Kapton and PPT films than in Novax. This qualitative comparison in crystallinity among films coincides with the quantitative evaluation of crystallinity by using galvanomagnetic properties, as will be discussed below.

The averaged grain size D, which corresponds to the average size of basal planes of single crystal domains in the films, was measured from these channeling contrast images.

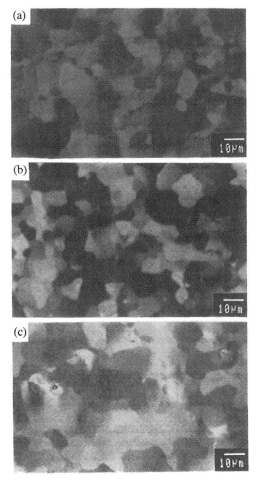

Figure 2-21 Electron channeling contrast images on the graphite films prepared from three polyimides (a) Kapton 3100, (b) PPT3200 and (c) Novax 3100 [Hishiyama, 1992].

The results from the present five films are listed in Table 2-3, in order to compare the mean free path determined from magnetoresistance.

In Fig. 2-22a and b, the transverse magnetoresistance $(\Delta\rho/\rho)_{max}$, measured at liquid nitrogen temperature (77 K) and liquid helium temperature (4.2 K), respectively, is plotted against the magnetic field B [Kaburagi et al., 1995].

The magnetic field dependences of $(\Delta\rho/\rho)_{max}$ at 77 K on these films are approximated by B^2, as reported on well-crystallized graphites. The values of $(\Delta\rho/\rho)_{max}$ for the films from Kapton and PPT are close to that for HOPG. Under high magnetic fields and at liquid helium temperature (4.2 K), $(\Delta\rho/\rho)_{max}$ reveals the Shubnikov-de Haas oscillation, as shown in Fig. 2-22b.

From the measurement of Hall coefficient R_H, the high crystallinity of the films from Kapton and PPT, comparable to HOPG, was also concluded [Kaburagi et al., 1995]. Its field dependences up to 1 T at 77 K and up to 6.5 T at 4.2 K are shown in Fig. 2-23a and b, respectively. In Fig. 2-23b, the Shubnikov-de Haas oscillation in R_H is observed, as in $(\Delta\rho/\rho)_{max}$ in Fig. 2-22b. Because the oscillation in R_H is more pronounced than in $(\Delta\rho/\rho)_{max}$, Landau levels for majority carriers can be assigned, as indicated by 3e and 3h (majority electrons and holes with $n = 3$, respectively), for example, in Fig. 2-23b.

In Table 2-3, the values of $(\Delta\rho/\rho)_{max}$ at 77 and 4.2 K under the magnetic field of 1 T, anisotropy ratio r, which is an average of r_{TL} and r_T, resistivity ratio $\rho_{300K}/\rho_{4.2K}$, and mean free path for carriers λ calculated from the value of $(\Delta\rho/\rho)_{max}$ at 4.2 K and 0.1 T are listed on five films, together with the mosaic spread MS measured by X-ray diffraction and average grain size D determined from electron channeling contrast images [Kaburagi et al., 1995].

The preferred orientation of graphite basal planes along the film surface is characterized by either mosaic spread MS or the magnetoresistance anisotropy ratio r, both showing that HOPG is superior to the other four films, but the films derived from polyimides are better than the film of PG3200 which has been heated up to 3200°C without pressure. The crystallinity in the films from Kapton and PPT, evaluated by either $\rho_{300K}/\rho_{4.2K}$, $(\Delta\rho/\rho)_{max}$ at 77 and 4.2 K, mean free path λ, or averaged grain size D, is comparable to HOPG and much better than PG3200. These results exhibit the fact that excellent crystallinity can be obtained from some polyimide films by simple heat treatment, i.e. without pressure, at high temperatures.

A correspondence between two parameters λ and D, which are determined by different techniques, is not quantitative; the D-value for HOPG is one order higher than those for other films, but the λ-value for HOPG is roughly twice those for others. This might suggest that the grains in HOPG consist of subgrains which have so close an orientation in the [11.0] direction as not to be detected by the electron channeling technique. The same phenomenon was observed in TEM where the mosaic domains containing Moiré fringes yielded values similar to D.

By compressing a pile of Kapton films with 25 μm thickness under 1.0–1.5 MPa during carbonization and then hot-pressing at 10–30 MPa up to a temperature of 2800–3000°C, highly oriented graphite blocks were prepared and commercialized for use as monochromators of X-rays and neutron beams [Murakami et al., 1992, 1993]. Some graphite blocks are shown in Fig. 2-24. Most characteristics of the graphite block thus

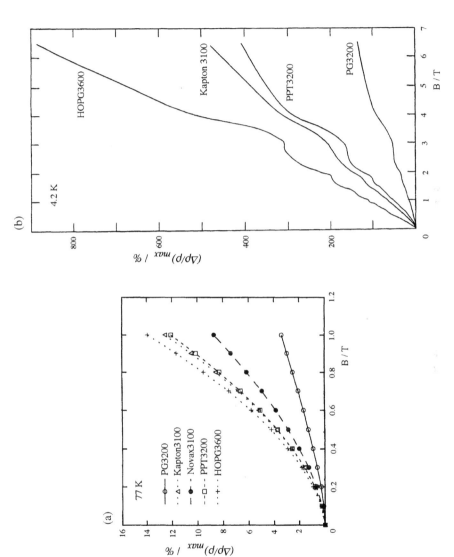

Figure 2-22 Dependences of maximum magnetoresistance $(\Delta\rho/\rho)_{max}$ on magnetic field strength B for highly oriented graphites from different precursors [Kaburagi et al., 1995].

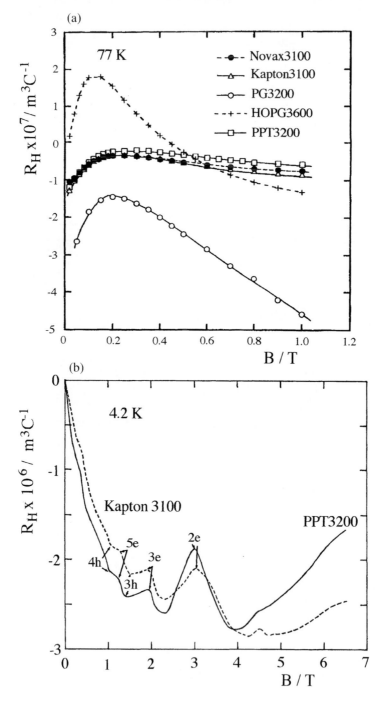

Figure 2-23 Dependences of the Hall coefficient R_H on magnetic field strength B for highly oriented graphites from different precursors [Kaburagi et al., 1995].

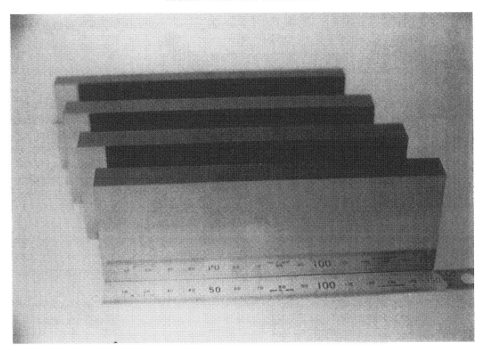

Figure 2-24 Highly oriented graphite blocks prepared from polyimide Kapton. [Courtesy of Dr Murakami of Matsushita Electric Co. Ltd.]

prepared were reported to be comparable with those of HOPG, in particular MS being about 0.4°.

CHAPTER 3

Isotropic Carbons

3.1. Realization of Isotropic Nature

The basic structural units (BSUs) in graphite-based carbon materials are hexagonal carbon layers, which are principally anisotropic because of different chemical bonds along and normal to the layer, i.e. covalent and van der Waals-like bonds, respectively. In the highly oriented graphites described in the previous chapter, these BSUs grow in a large size along and across the layers, in which the stacking of layers is regular, ABAB.... As a consequence, high orientation, strong anisotropy and high crystallinity result, even though the size of the materials is rather limited to either thin films or plates. Because of various characteristics of carbon materials, however, there is a wide range of applications for carbon materials with large sizes and various shapes where the isotropic nature is essentially required to the bulk of carbon materials.

In order to realize the isotropy in carbon materials which fundamentally consist of anisotropic BSUs, two methods have been employed: random aggregations of micrometer- or millimeter-sized graphitic crystallites in the bulk and of nanometer-sized hexagonal layers accompanying a large number of crossed pores [Inagaki and Hishiyama, 1994; Oya, 1996, 1997a].

The former is represented by isotropic high-density graphites where a certain size of hexagonal carbon layers is desired in order to maintain high electrical and thermal conductivity, as well as high thermal shock resistance. Optical micrographs of the polished cross-section of these graphites with different bulk densities are shown in Fig. 3-1. The mechanical properties, elastic modulus, bending strength, fracture toughness and critical crack size are summarized in Table 3-1 for each graphite, together with pore structure, bulk density, porosity and averaged area of pores obtained by image analysis on these optical micrographs [Oshida et al., 1996]. In these graphite blocks, pores are distributed homogeneously with relatively homogeneous size. With the increase in bulk density, i.e. the decrease in porosity and average pore area, mechanical properties were improved, with elastic modulus, bending strength and fracture toughness also increasing

The latter, i.e. aggregation of nanometer-size carbon layers, was realized in glass-like carbons [Noda et al., 1969]. Figure 3-2 illustrates a structural model of these glass-like carbons [Shiraishi, 1984], where the glass-like carbon is postulated to be heat-treated at 3000°C. The structure of these carbons was also expressed by a model of a crumpled sheet

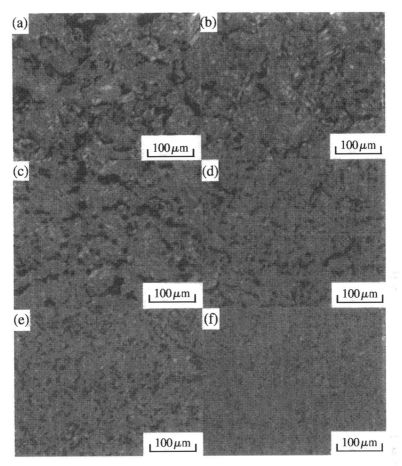

Figure 3-1   Optical micrographs of polished surfaces of isotropic high-density graphites [Oshida et al., 1996].

Table 3-1. Pore structure in isotropic high-density graphites obtained by image analysis and their mechanical properties [Oshida et al., 1996]

Sample	Bulk density (g cm^{-3})	Porosity (%)	Area of pores (μm^2)	Elastic modulus (MPa)	Bending strength (MPa)	Fracture toughness (MPa/m$^{1/2}$)	Critical crack size (mm)
A	1.735	21.9	545	882	25.3	0.63	0.064
B	1.788	25.3	506	1026	39.7	0.80	0.067
C	1.842	21.1	395	1123	52.2	0.99	0.072
D	1.842	21.7	244	1122	48.0	0.96	0.071
E	1.848	20.9	155	1232	70.9	1.04	0.068
F	1.802	12.0	31	1301	91.4	0.87	0.046

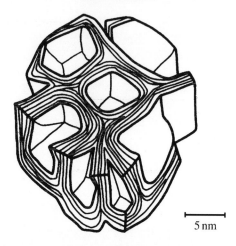

5 nm

Figure 3-2 Structural model for glass-like carbon after heat treatment at temperatures as high as 3000°C [Shiraishi, 1984].

of carbon hexagons [de Fonton et al., 1980; Oberlin, 1989]. It is characteristic for these carbon materials that only few nanometer-sized layers construct shells even after 3000°C treatment, this nanotexture resulting in the isotropic and non-graphitizing nature of these carbons in bulk [Inagaki, 1985b].

3.2. Isotropic High-density Graphites

(a) Production

The production procedure is shown as a block diagram in Fig. 3-3, which is classified on the basis of the raw materials used and either the use of a binder or not. The fundamental procedure for method A is the same as the conventional one which has been applied to graphite electrodes, using the pastes of filler coke particles with binder pitch followed by forming, carbonization and graphitization at high temperatures [Inagaki, 1999]. Key techniques to produce isotropic high-density graphite are the use of fine particles of filler cokes and cold isostatic pressing (CIP) for forming.

In Fig. 3-4, a principle of CIP is illustrated. By applying this forming process, various advantages in production were obtained; it became possible to use fine particles of filler cokes (e.g. about 5 μm), to attain high homogeneity in the products because of good conductance of pressure, to obtain near-shaped products, and to achieve a high density of the final products by using a relatively large amount of binder pitch, etc. However, very accurate control during the carbonization and graphitization process is required in order to avoid deformation and cracking of the blocks, e.g. selection of raw materials, filler cokes and binder pitches, pressurizing program during CIP and temperature control during carbonization.

The detailed conditions for the production of isotropic high-density graphites are not published because they are highly confidential issues for the industries. Therefore, rather old data could be referred here to understand the importance of some of production parameters. In Fig. 3-5a, bulk density of the carbon blocks before and after carbonization was plotted against the content of binder pitch [Okada and Takeuchi, 1960]. This result suggests that there is an optimum content of binder pitch, and the decrease in density during carbonization is also related to binder content.

To select the binder giving a high carbon yield after carbonization is also important for the densification of the graphite blocks. Carbon yield from pitch was known to depend on the pitch itself and on the preparation conditions for binders, but also on the content of filler cokes. Figure 3-5b shows the carbon yield of a binder pitch as a function of content of filler cokes with particle size less than 40 µm, revealing its strong dependence

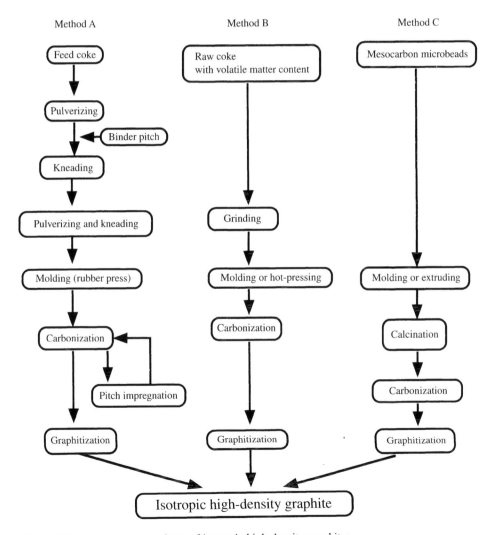

Figure 3-3   Production procedures of isotropic high-density graphites.

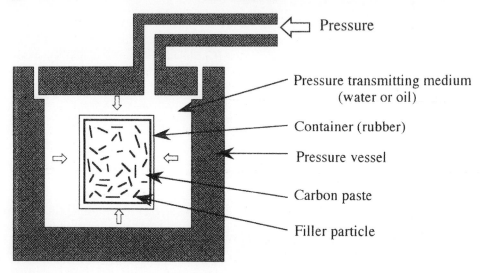

Figure 3-4   Principle of cold isostatic pressing. Anisotropic crystallites are shown by bars.

[Gilorami, 1963]. If filler content becomes larger than a certain amount, binder pitch forms a thin film on the surface of the filler particles because of its good wettability and its flow at high temperatures during carbonization is supposed to be disturbed. In order to achieve a high carbon yield, there were some trials using additives for binder pitch, but this was not accepted as an industrial process because it lowered the purity of the final products, high purity of these graphite blocks being an important characteristic for certain applications.

The formed carbon pastes can be carbonized under pressure [Matsuo et al., 1968]. In Table 3-2, the characteristics of graphite obtained by carbonization of the formed carbon paste with the bulk density of 1.67 g cm^{-3} under argon pressure of 5 MPa is compared to that calcined under normal pressure. By pressure carbonization up to 600°C, bulk density, bending strength, Young's modulus and electrical conductivity of the block show a marked increase.

In method B in Fig. 3-3, raw coke powder was used as a starting material, which was manufactured at about 500°C, contained about 8.5 wt% volatile matter and sintered without any binder [Ogawa et al., 1981]. An advantage of this method is self-sintering of raw coke; in other words, no binder is necessary, but severe control of the carbonization process is required. This method is now used to produce the carbon/metal carbide composites in industry, which will be described in Chapter 7.

The use of spherical particles of filler was proposed as an alternative production process (method C in Fig. 3-3) for isotropic high-density graphite. Mesophase spheres, spheres with an optically anisotropic texture formed at the very beginning of pyrolysis of pitch, which were separated from isotropic matrix pitch (mesocarbon microbeads, MCMB, Fig. 3-6) [Honda et al., 1973; Yamada et al., 1974], were formed into a block and then carbonized. Properties of three commercially available mesocarbon microbeads are compared in Table 3-3; their particle size and ash content are different in the products.

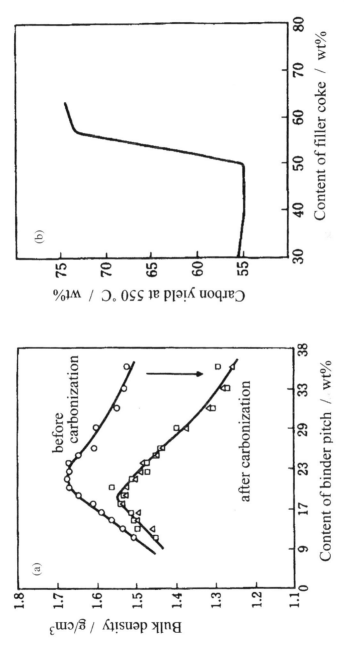

Figure 3-5 Bulk density of carbon blocks after carbonization as a function of the content of binder pitch in carbon paste [Okada and Takeuchi, 1960] (a) and carbon yield from binder pitch as a function of the content of filler coke [Gilorami, 1963] (b).

Table 3-2. Properties of graphite blocks carbonized with or without pressure [Matsuo et al., 1968]

Carbonization condition	Under pressure (5 MPa, 600°C)	Under atmospheric pressure (800°C)
Bulk density (g cm^{-3})	1.686	1.471
Electrical resistivity ($\times 10^{-6}$ Ω cm)	79	106
Bending strength (MPa)	26.6	14.6
Young's modulus (MPa)	111	69
Shore hardness	41	32
Thermal expansion coefficient (75–375°C, $\times 10^{-6}$ °C^{-1})	2.39	2.40

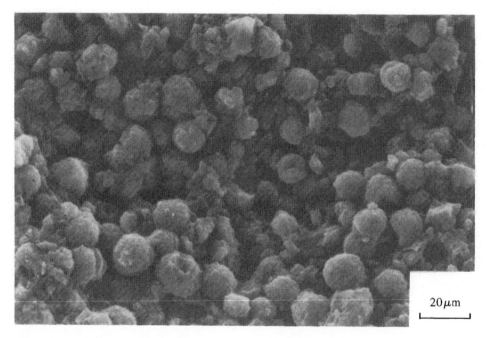

$20\mu m$

Figure 3-6   Mesocarbon microbeads. [Courtesy of Kawasaki Steel Co. Ltd.]

Table 3-3. Characteristics of commercially available mesocarbon microbeads

Company	Kawasaki Steel	NKK	Osaka Gas
Toluene-insoluble (wt%)	97.0–99.5	–	94–98
Quinoline-insoluble (wt%)	92.0–96.0	95.9–97.0	80–93
Fixed carbon (wt%)	–	85.0–89.0	87–90
Volatile matter (wt%)	0.5–9.0	–	10–13
Ash content (ppm)	400	Trace	100–300
Particle size (μm)	13–17	5–50 (av. 10)	D_{10}: 2–3 D_{50}: 5–7 D_{90}: 9–15

Using method C to produce isotropic high-density graphites has some advantages; isotropy in bulk is obtained even by using molding because the raw materials are spheres, although forming by CIP is recommended in practice, in order to avoid possible inhomogeneity in the formed blocks; no binder pitch is necessary because the spheres still contain a relatively large amount of volatile matter which can act as a binder during carbonization. However, there are also some disadvantages, including the low yield of these spheres from pitch and the difficulty in separating them from the matrix pitch, both of which lead to increased production costs.

(b) Properties and applications

Some characteristics of commercially available isotropic high-density graphites are tabulated in Table 3-4. On these graphites the difference in various properties along different directions of the blocks is less than 3%. These graphites were firstly industrialized for the applications of hot-pressing molds, nozzles for continuous casting of metals and electrodes for electrical discharge machining (EDM), but now have wide applications for crystal growth of semiconductor crystals, such as silicon, germanium, I-V and II-VI semiconductors, as heaters and crucibles. There are many additional applications, such as boats and assemblies for liquid epitaxy, susceptors and wafer trays for different CVD processes, shields, and electrodes and ion sources for ion implantation. Some of the products are shown in Fig. 3-7. These applications in the field of semiconductor industries not only increased the production but also improved markedly

Table 3-4. Characteristics of commercial isotropic high-density graphites

Company	Brand name	Bulk density $(g\ cm^{-3})$	Electrical resistivity $(\mu\Omega\ m)$	Bending strength (MPa)	Shore hardness	Thermal expansion coefficient $(\times 10^{-6}\,°C^{-1})$
POCO	AXF-50	1.83	1480	91	75	8.4
	ZXF-50	1.83	1580	127	88	7.7
R/W	EK 92	1.75	1100–1500	30–35	40–50	2.8
	EK 95	1.83	1300–1900	35–40	55–65	2.8
LCL	P 7477	1.75	1600	50	–	5.5
	1116	1.82	1500	65	65	5.5
AGL	1M1	1.77	900	23	–	4.6
	1M2	1.82	1000	23	–	4.6
Stackpole	2020	1.77	2030	30	50	4.2
	2019	1.82	1900	42	70	4.7
Ibiden	ET-10	1.75	1300	150	50	3.5
	T-6	1.92	1600	100	85	6.5
Tokai	G 345	1.76	1100	42	50	3.7
Carbon	G 548	1.85	1200	70	68	4.5
Toshiba	IP 40	1.70	1200	30	45	3.8
Ceramics	NP 60	1.80	1400	45	50	4.5
Toyo	IG-11	1.77	1100	40	54	4.6
Tanso	ISO-88	1.90	1500	95	90	6.5
Nihon	EGM-63	1.78	1500	38	50	4.8
Carbon	EGM-74	1.82	1450	40	55	4.8
Hitachi	PD-320	1.78	1100	47	47	5.2
Kasei	PD-610	1.83	1100	55	65	6.0

the quality of these graphites. They are used because of their accurate and easy machining, high strength, isotropy in electrical resistivity and high purity. The demand due to a marked increase in integration in large-scale integration (LSI) devices requires high finishing of the surface of graphite wafers: "on the surface of the Tokyo dome for professional baseball games even a sand grain of 0.1 mm is not allowed".

These isotropic high-density graphites were also used for the structural parts of many fission reactors. In Fig. 3-8, the construction of a high-temperature gas-cooled test reactor in Japan is shown, where graphites are used as permanent reflectors at the outer part, replaceable reflectors at the inner part and fuel-element blocks at the center. For these nuclear applications, graphite materials have to have high purity, high strength and isotropic nature. Specifications for graphite materials used for Japanese test reactors are summarized in Table 3-5. In relation to this application, the detailed mechanical performance of these isotropic high-density graphites, mechanical properties under complicated loads, different temperatures and different atmospheres, very detailed fatigue tests, etc., have been studied [Eto et al., 1998]

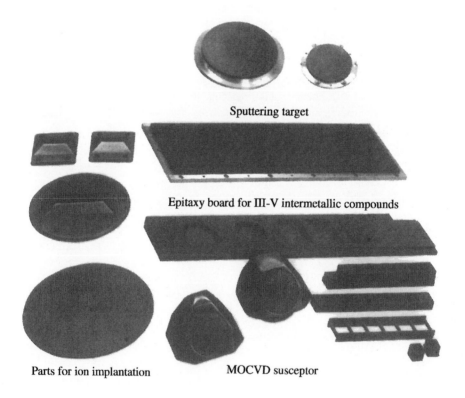

Sputtering target

Epitaxy board for III-V intermetallic compounds

Parts for ion implantation   MOCVD susceptor

Figure 3-7 Industrial products of isotropic high-density graphite for semiconductor industries. [Courtesy of Tokai Carbon Co. Ltd.]

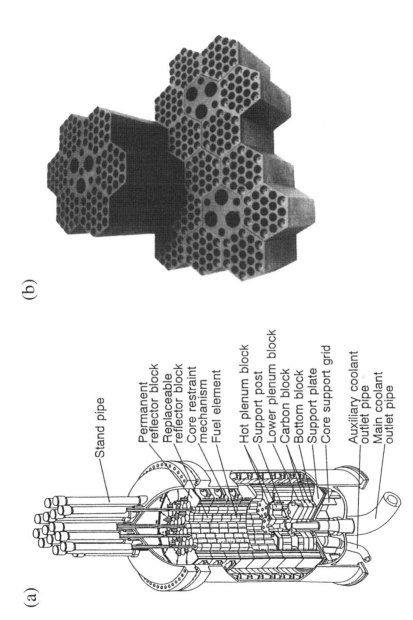

Figure 3-8 Construction of Japanese high-temperature gas-cooled reactor (a) and replaceable reflector blocks of isotropic high-density graphite (b). [Courtesy of Japan Atomic Energy Research Institute.]

Table 3-5. Specifications for graphite materials for Japanese high-temperature gas-cooled test nuclear reactors [Courtesy of Japan Atomic Energy Research Institute]

	Replaceable reflectors & fuel-element blocks	Permanent reflectors
Bulk density (g cm^{-3})	1.78	1.73
Tensile strength (MPa)	25.8	8.3[a]
Compressive strength (MPa)	78.4	32.1[a]
Thermal expansion coefficient ($\times 10^{-6}$ °C^{-1})	4.06	2.34[a]
		2.87[b]
Thermal conductivity (cal cm^{-1} s^{-1} °C^{-1})	0.19	0.18[a]
Ash content (ppm)	<100	<7000
Grain size (μm)	20	<800

[a] Radial direction; [b] axial direction.

These graphite materials are also used in fusion reactors as interior liners, movable limiters, diverters and specialized fixtures. The inner structure of Japanese test reactor JT-60 is shown in Fig. 3-9. The low atomic number of graphite materials is the most important factor in reducing the interference with plasma, but many problems for practical use, such as resistance to physical and chemical sputtering, thermal shock resistance and recycling of hydrogen, have to be solved.

3.3. Glass-like Carbons

(a) Production

Glass-like carbons are characterized by very broad lines of X-ray diffraction, because of which they have been called "amorphous" carbons, although the basic structural units are known to be small hexagonal carbon layers, and have really glass-like characteristics, such as high hardness, brittle conchoidal fracture and low gas permeability [Noda et al., 1969]. They are produced by pyrolysis of thermosetting resins, such as phenol-formaldehyde and polyfurfuryl alcohol, and also of cellulose under an exactly controlled heating process. The production process is illustrated in Fig. 3-10. Formation of glass-like carbon is a very slow process mostly limited by the rate at which decomposition gases evolve out from the precursor blocks without leaving open pores behind [Fitzer et al., 1969]. They are very hard and brittle, and their machining is so difficult that they have to have a shape near to that of the final product before carbonization. Therefore, their carbonization process has to be controlled exactly by taking account of rather large shrinkage during pyrolysis and carbonization.

In Table 3-6, some commercially available glass-like carbons are listed along with their properties. All of them have very low gas permeability, 10^{-12} to 10^{-7} cm^3 s^{-1}, and also very low porosity, even though their bulk density is rather low, 1.4–1.5 g cm^{-3}, indicating the presence of a large amount of closed pores. This gas impermeability is retained even after high-temperature heat treatment up to 3000°C.

Figure 3-9 Inner structure of Japanese test fusion reactor JT-60. [Courtesy of Japan Atomic Energy Research Institute.]

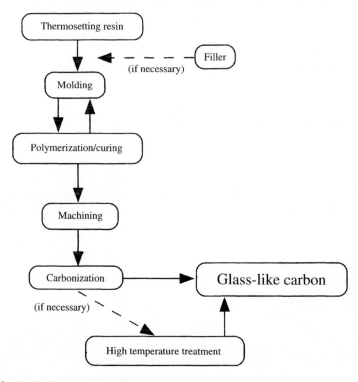

Figure 3-10 Production procedures for glass-like carbons.

Table 3-6. Characteristics of commercially available glass-like carbons

Company	Tokai Carbon			Kao
Brand name	GC–10	GC–20	GC–30	S–100
HTT (°C)	1300	2000	3000	1200
Bulk density (g cm^{-3})	1.47–1.51	1.46–1.50	1.43–1.47	1.45
Gas permeability (cm^2 s^{-1})	10^{-11}–10^{-12}	10^{-11}–10^{-12}	10^{-7}–10^{-9}	
Bending strength (MPa)	88–98	98–118	49–59	98
Young's modulus (GPa)	29–32	29–32	22–25	20
Shore hardness	100–120	100–110	70–80	120
Electrical resistivity (μΩm)	45–65	40–45	35–40	45
Thermal conductivity (Ωm^{-1} K^{-1})	3.76–4.6	8.36–9.19	15.0–17.6	3.34–3.76
Thermal expansion coefficient ($\times 10^{-6}$ °C^{-1})	2.0–2.2	2.0–2.2	2.0–2.2	3.5

(b) Structure and texture

One characteristic of glass-like carbons is their isotropic nanotexture. As shown in Fig. 3-2, they are composed of fundamental units of small pores, the walls of which are composed of minute hexagonal carbon layers, basic structural units (BSUs). Most of these pores are closed, this being the reason why they are gas impermeable despite being of low bulk density. Such a closed pore structure is formed in localized regions of most carbons heat-treated at low temperatures, particularly carbons prepared from thermosetting resins. In the present book, therefore, glass-like carbons are defined as carbon blocks with a controlled or designed shape, not small particles with an irregular shape or powder, with disordered structure, gas impermeability and high hardness. In order to achieve these characteristics, exact control of the production process is necessary. The term "non-graphitizing" reveals the nature of carbon materials which do not show appreciable development of a graphitic structure even by high-temperature treatments. Therefore, all glass-like carbons discussed here are non-graphitizing and belong to non-graphitizing, or hard, carbons. So, the definition of glass-like carbons in this book is the non-graphitizing carbons produced into designed shapes.

Micrographs taken by a scanning electron microscope with field emission source (FE-SEM) on the fractured surface of different glass-like carbons are shown in Fig. 3-11 [Yoshida et al., 1991]. All samples show a granular texture, which agrees with the structural model shown in Fig. 3-2. The size of constituent grains is measured from these micrographs for each sample and listed in Table 3-7, together with X-ray and magnetoresistance parameters measured on the same samples. The average grain size in glass-like carbon increases, albeit slightly, with the increase in heat treatment temperature, associated with a small increase in crystallite sizes, $L_c(002)$ and $L_a(110)$, and a slight decrease in d_{002}. On two samples, both of which are supposed to be treated at a low temperature around 1000°C, no magnetoresistance was measured even at a magnetic field strength as high as 6.5 T. Anisotropy ratios r_T and r_{TL} are close to 1, indicating isotropy in structure.

For non-graphitizing or hard carbons, a 002 lattice fringe image as shown in Fig. 3-12a is usually obtained after heat treatment at temperatures as high as 2800°C [de Fonton et al., 1980; Oberlin, 1989]. From such lattice fringe images, a ribbon model has been proposed [Jenkins and Kawamura, 1971]. Through detailed studies using high-resolution TEM techniques, such as selected area electron diffraction (Fig. 3-12b) and 110 dark-field images (Fig. 3-12c), a structure model represented by closed pores with walls of locally oriented distorted carbon layers has been proposed, as shown in Fig. 3-13 [Oberlin, 1989]. As shown by the 110 dark-field image in Fig. 3-12c, the regions surrounded by 002 planes in lattice fringes (Fig. 3-12a) are lightened, revealing that 002 layers are perpendicular to the incident electron beams, lying down in the micrograph. Therefore, it seems more reasonable to postulate pores with walls consisting of 002 layers, i.e. the model of crumpled sheets in Fig. 3-13 [Oberlin, 1989], than to assume the presence of the ribbons of stacked 002 layers.

From TEM observations on 2500°C-treated non-graphitizing carbon, a pore diameter of 3–6 nm and pore wall thickness of 2–4 nm were measured, indicating that the grain size, i.e. the size of grains in the model with locally oriented walls, observed on fractured surface may be about 7–10 nm. These TEM observations agree well with FE-SEM

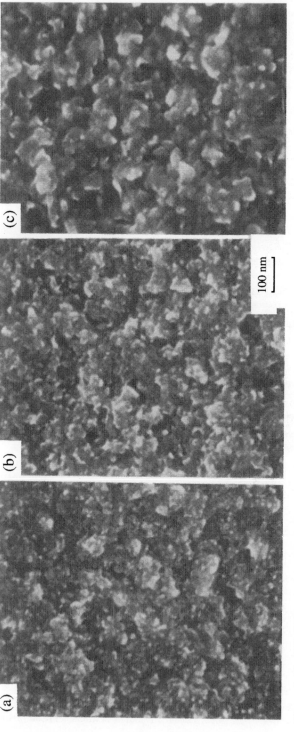

Figure 3-11 FE-SEM micrographs of the fractured surface of glass-like carbons [Yoshida et al., 1991]. (a) 1000°C-treated (GC-10), (b) 2000°C-treated (GC-20) and (c) 3000°C-treated (GC-30).

Table 3-7. Grain size, D, determined from FE-SEM, X-ray parameters d_{002}, $L_c(002)$ and $L_a(110)$, and magnetoresistance parameters $(\Delta\rho/\rho)_{cr}$, r_T and r_{TL} on glass-like carbons [Yoshida et al., 1991]

Sample	D (nm)	d_{002} (nm)	$L_c(002)$ (nm)	$L_a(110)$ (nm)	$(\Delta\rho/\rho)_{cr}$ (%)	r_T	r_{TL}
GC-10	7.0	0.3468	1.9	2.5	−	−	−
GC-20	10.0	0.3442	3.3	3.1	− 0.0846	0.771	0.887
GC-30	13.1	0.3436	3.6	3.5	− 0.1821	0.962	0.860
AGC	6.0	0.3470	1.9	2.0	−	−	−
UDAC	9.4	0.3443	3.7	3.0	− 0.0696	1.00	1.00

observation on various glass-like carbons summarized in Table 3-7. The pore wall thickness determined from TEM observation corresponds roughly to the $L_c(002)$ value determined by X-ray diffraction (XRD) on glass-like carbons. Therefore, bulk density was calculated by assuming the grain of diameter D with wall thickness of $L_c(002)/2$ and using observed d_{002} and a_0, which showed a rough agreement with the measured bulk density on each sample [Yoshida et al., 1991]. Therefore, the structure model consisting of closed pores shown for non-graphitizing carbons in Fig. 3-2 is exactly the same as the one shown in Fig. 3-13, with just a slightly different visualization of the structure. The ribbon model for non-graphitizing carbons seems to be ineffective for interpreting the structure, texture

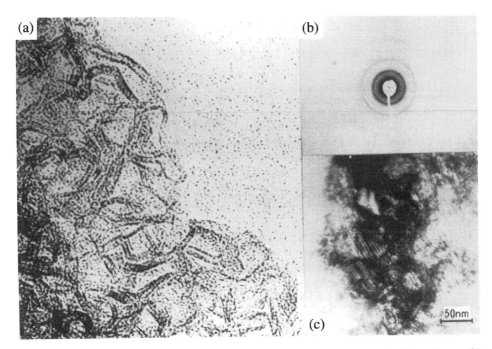

Figure 3-12 TEM micrographs on 2500°C-treated sugar coke. (a) 002 lattice fringe image, (b) selected area electron diffraction pattern and (c) 110 dark-field image. [Courtesy of Mme Oberlin.]

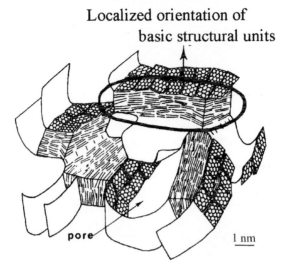

Figure 3-13   A model for hard carbons. [Courtesy of Mme A. Oberlin.]

and properties of glass-like carbons, as it is difficult to explain the presence of a large number of closed pores and gas impermeability.

c) Properties

Owing to the non-graphitizing nature of glass-like carbons, their electromagnetic and mechanical properties are quite different from the highly oriented graphites described in

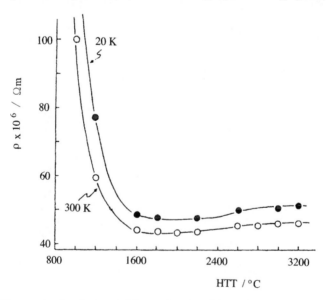

Figure 3-14   Changes in electrical resistivity ρ of a glass-like carbon at 20 and 300 K with HTT [Yamaguchi, 1963a].

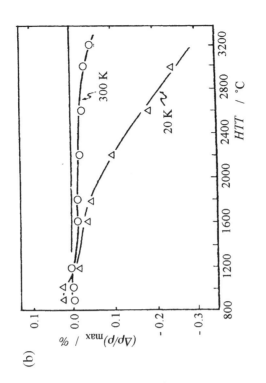

(b)

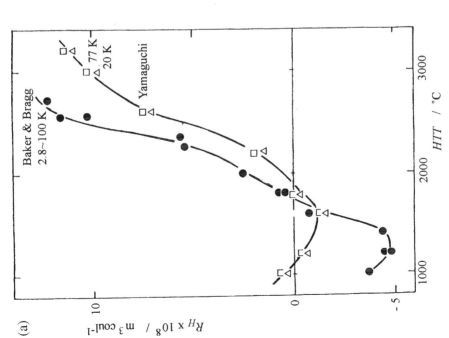

(a)

Figure 3-15 Changes in Hall coefficient R_H (a) and maximum magnetoresistance $(\Delta\rho/\rho)_{max}$ (b) of a glass-like carbon at 20 and 300 K with HTT [Yamaguchi, 1963b].

the previous chapter and isotropic high-density graphites in the last section of the present chapter.

The electrical resistivity ρ of a glass-like carbon measured at 20 and 300 K is plotted against HTT in Fig. 3-14 [Yamaguchi, 1963a]. With increasing HTT, ρ decreases at first, passes through a shallow minimum and then increases gradually. The HTT dependences of the Hall coefficient R_H and maximum transverse magnetoresistance $(\Delta\rho/\rho)_{max}$ measured for two glass-like carbons at different temperatures in a magnetic field of 0.65 T are shown in Fig. 3-15a and b, respectively [Yamaguchi, 1963b; Baker and Bragg, 1983]. Even after heat treatment at 3200°C, glass-like carbons exhibit positive R_H and negative $(\Delta\rho/\rho)_{max}$, which are characteristic for non-graphitizing carbons. R_H for glass-like carbons is

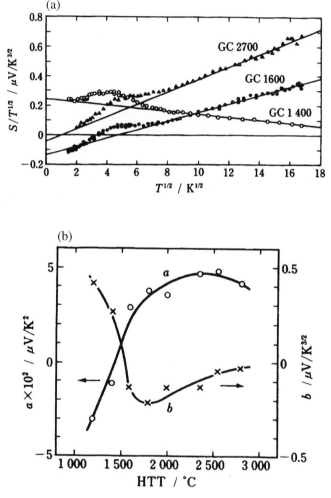

Figure 3-16 Temperature dependences of thermoelectric power S for glass-like carbons with different HTT (a) and changes in constants a and b in temperature dependence of S with HTT (b). [Courtesy of Profs Hishiyama and Kaburagi of Musashi Institute of Technology.]

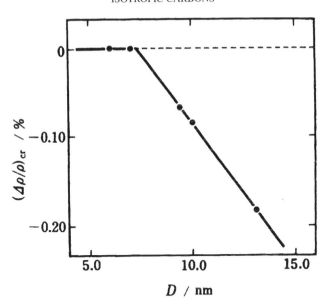

Figure 3-17 Crystallite magnetoresistance $(\Delta\rho/\rho)_{cr}$ of glass-like carbons shown in Fig. 3-11 and Table 3-7 against average grain size D determined by FE-SEM [Yoshida et al., 1991].

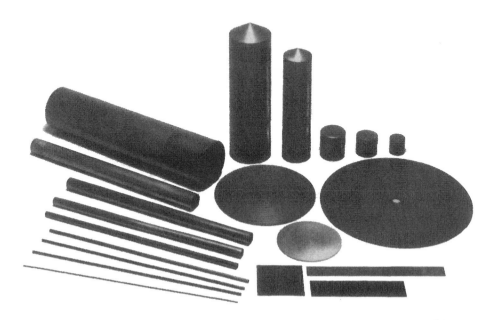

Figure 3-18 Industrial products of glass-like carbon. [Courtesy of Tokai Carbon Co. Ltd.]

independent of magnetic field and temperature of measurement in a range of 2.8–300 K. However, it shows a strong dependence on HTT and a change from negative to positive values around 1600–1700°C. The dependence of thermoelectric power S on the temperature of measurement T was determined on glass-like carbons with different HTT. The results are shown in Fig. 3-16a by plotting $S/T^{1/2}$ against $T^{1/2}$ [Kaburagi et al., 1986; Hishiyama and Kaburagi, 1987]. The dependence of S on T can empirically be written as:

$$S = aT + bT^{1/2} + S_B,$$

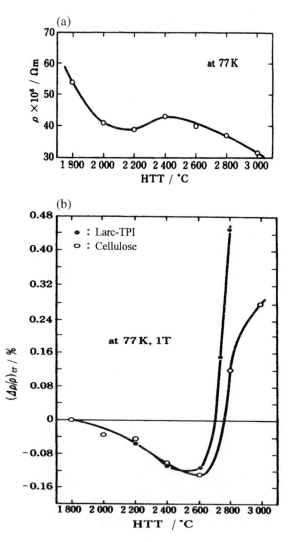

Figure 3-19 HTT dependence of resistivity ρ (a) and crystallite magnetoresistance $(\Delta\rho/\rho)_{cr}$ of carbon films with glass-like carbon structure prepared from polyimide film Larc-TPI and cellulose film [Hishiyama et al., 1993b].

where the first term is a strong-scattering metallic component, the second is attributed to variable-range hopping and the third is a peaked component. The signs of the constants a and b give those of the majority carriers for metallic and hopping conduction, respectively. HTT dependences of these two constants on a glass-like carbon are shown in Fig. 3-16b [Hishiyama and Kaburagi, 1987]. The majority carriers for metallic conduction in glass-like carbons are changed from negative electron to positive holes around 1500°C, which shows a good correspondence to the change in R_H in Fig. 3-15a. The sign of majority carriers for variable-range hopping is changed at the same temperature from positive to negative, but its contribution to the observed value of thermoelectric power becomes small with the increase in HTT.

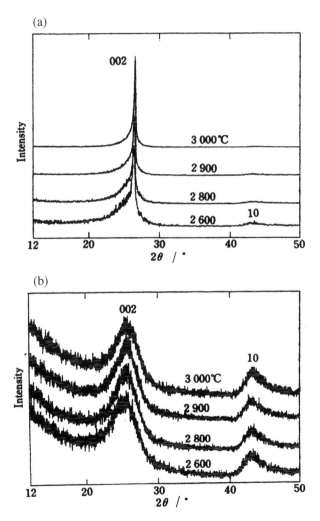

Figure 3-20   X-ray diffraction patterns measured by reflection and transmission modes (a and b, respectively) on cellulose-derived carbon films with different HTT [Hishiyama et al., 1993b].

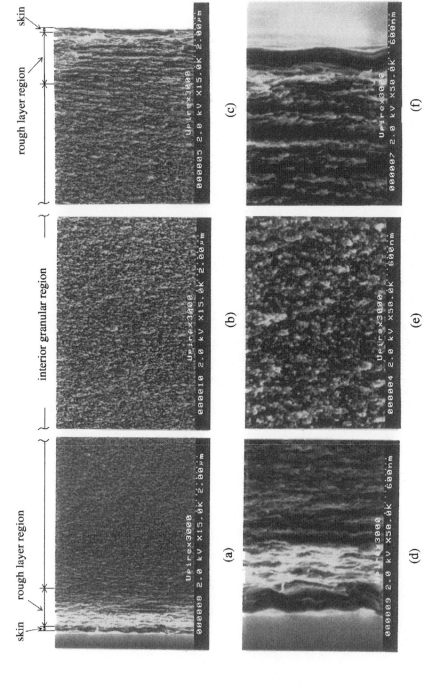

Figure 3-21   SEM micrographs of the cross-section of the polyimide film Upilex after the heat treatment at 3000°C [Hishiyama et al., 1998].

In Fig. 3-17, crystallite transverse magnetoresistance of the glass-like carbons, used for FE-SEM observation and XRD (Fig. 3-11 and Table 3-7), is plotted against average grain size D determined from SEM micrographs. These results suggest a strong dependence of galvanomagnetic properties on structure, as with all other carbon materials.

Some industrial products of glass-like carbon are shown in Fig. 3-18. Not only thin plate but also pipes, rods and crucibles are produced and used in various industrial fields.

Several authors have pointed out that on different glass-like carbons there is a thin surface layer with better crystallinity than on the inside, particularly on those heat-treated at high temperatures. This phenomenon was observed markedly on thin films derived from thin cellulose film and a polyimide film Larc-TPI, and studied in detail by measuring different properties [Hishiyama and Kaburagi, 1987; Hishiyama et al., 1993b] In Fig. 3-19a and b, electrical resistivity ρ and crystallite magnetoresistance $(\Delta\rho/\rho)_{cr}$ at 77 K under 1 T are plotted against HTT on these two glass-like carbon films, respectively. By comparison of Fig. 3-19a and b with Figs 3-14 and 3-15b, respectively, the behaviors of ρ and $(\Delta\rho/\rho)_{cr}$ on these films are very similar to other glass-like carbons up to HTT of 2600°C. Above this temperature, however, their HTT dependences are quite different; ρ decreases gradually with the increase in HTT and $(\Delta\rho/\rho)_{cr}$ increases abruptly to become positive (Fig. 3-19a and b, respectively), instead of a gradual increase in ρ and a continuous decrease in negative $\Delta\rho/\rho$ in other glass-like carbons (Figs 3-14 and 3-15b, respectively). This has been proven to be due to the formation of thin graphitic layers on the surface of films from XRD and FE-SEM studies.

In Fig. 3-20, XRD patterns measured by reflection and transmission arrangements of the film specimen to an incident X-ray beam are shown on cellulose-derived films heat-treated above 2600°C [Hishiyama and Kaburagi, 1989]. The 002 profile measured by reflection mode shows a sharp peak overlapping a broad one, but that measured by transmission mode only a broad peak, suggesting the formation of well-crystallized regions by orienting along the film surface. The scanning electron micrograph of the cross-section of the film, which is derived from a polyimide Upilex, shown in Fig. 3-21 reveals that these well-graphitized and well-oriented regions are formed on the surface of the film [Hishiyama et al., 1998].

CHAPTER 4

Carbon Fibers

4.1. Classification of Carbon Fibers

Carbon fibers (Fig. 4-1) which have been produced on either an industrial or a pilot scale are classified by the precursors used, as shown in Table 4-1 together with some characteristics of each fiber. The PAN-, isotropic-pitch- and mesophase-pitch-based carbon fibers are produced by the spinning of each precursor, polyacrylonitrile (PAN), isotropic pitch and anisotropic mesophase pitch, respectively, followed by stabilization and carbonization up to about 1300°C. In contrast to these carbon fibers derived from organic precursor fibers, vapor-grown carbon fibers are prepared by thermal decomposition of vapor of organic precursors, where stabilization is not needed. The first three carbon fibers, PAN- and two pitch-based, can be obtained as a continuous fibers and different morphologies as monofilaments, strands of 1000–12,000 filaments (Fig. 4-1a), cloth (woven fabric, Fig. 4-1b) with different woven modes, chopped fibers and also non-woven mats (felt and paper, Fig. 4-1c). Vapor-grown carbon fibers, however, are only short, but have a wide range of length. From the viewpoint of crystallinity, which can be attained by high-temperature treatment, i.e. graphitizability; however, vapor-grown carbon fibers have a great advantage over the other three. Some mesophase-pitch-based carbon fibers with a ribbon-like cross-section were reported to have a high degree of graphitization [Edie et al., 1994].

The carbon fibers are also classified according to their mechanical performance, tensile strength and modulus, because their principal application is the mechanical reinforcement of various composites. Figure 4-2 shows the classification based on mechanical performance.

Carbon fibers which have relatively low tensile strength and modulus, around 1000 MPa and 100 GPa, respectively, are classified into general-purpose grade (GP-grade). Isotropic-pitch-based and some PAN-based carbon fibers belong to this grade and are used in applications that benefit from their low weight and bulkiness, e.g. thermal insulation for a high-temperature furnace. A recent, novel application of these GP-grade carbon fibers is in cement reinforcement, which will be described in Chapter 7.

Carbon fibers with higher strength and modulus than GP-grade are called high-performance grade (HP-grade), which are further classified into high-strength type (HT type) and high-modulus type (HM type). Most PAN-based carbon fibers produced in

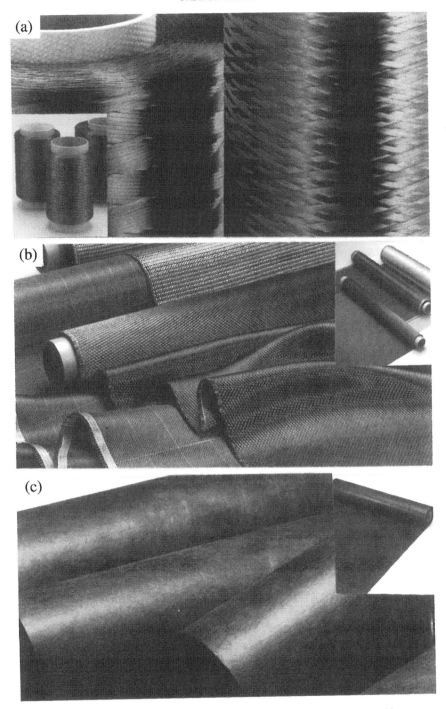

Figure 4-1   Carbon fibers: (a) strand, (b) cloth, (c) mat. [Courtesy of Tray Co. Ltd.]

Table 4-1. Classification of carbon fibers and their characteristics

Precursor	Carbon fibers	Characteristics
Polyacrylonitrile (PAN)	PAN-based carbon fibers	Different grades and types
Isotropic pitch	Isotropic pitch-based carbon fibers	GP grade
Anisotropic mesophase pitch	Mesophase pitch-based carbon fibers	HP grade, HM types Various textures in cross-sections
Hydrocarbon gases	Vapor-grown carbon fibers	High graphitizability Annual-ring texture in cross-section

industry are HT type, and mesophase-pitch-based carbon fibers which are produced through spinning of anisotropic pitches are HM type. In the production of PAN-based carbon fibers, great effort has been paid to obtain a high modulus and some of them could have a relatively high modulus. In contrast, in mesophase-pitch-based carbon fibers, the

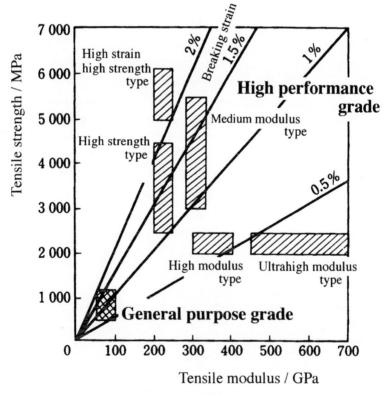

Figure 4-2   Classification of carbon fibers on the basis of mechanical performance.

aim is to achieve a high strength; some commercialized fibers have a relatively high strength. Detailed scientific aspects of carbon fibers, particularly PAN-based fibers, have been reviewed previously [Bunsell, 1988; Donnet et al., 1998].

In the present chapter, an overview of the production, structure and properties related to applications of carbon fibers will be presented. Therefore, no emphasis of precursors for these carbon fibers was afforded, although production and structure were discussed on the basis of precursors, because they governed strongly both the production procedure and the resultant structure of carbon fibers.

4.2. Production

(a) PAN-based carbon fibers

The production process is shown as a flow chart in Fig. 4-3a. It consists of the spinning of the precursor PAN, stabilization of spun fibers and heat treatment of stabilized fibers to high temperatures. It has been pointed out that the key technology for the production of PAN-based carbon fibers is the design of the precursor and the selection of its stabilization condition. Because of their confidential aspects for industry, however, the practice of these key technologies has never been published.

In order to have high spinnability of precursor PAN, various additives, including those which have been used for the production of PAN fibers themselves for clothes, have been tried. For the production of carbon fibers, not only high spinnability but also high rate stabilization and high mechanical performance of the resultant carbon fibers have to be attained, because their high producibility, which is directly related to the cost performance of carbon fibers, is demanded. Sodium acrylate comonomer was recently reported to be effective in halving the stabilization time compared with conventional procedures and also having a high strength of carbon fibers [Ogawa, 1994a; Ogawa and Saito, 1995].

Structural changes from precursor PAN to carbon during stabilization and carbonization have been studied by various authors [Bahl, 1998]. The changes in molecular structure with heat treatment to high temperatures are illustrated in Fig. 4-4; the cyclization at carboxyl radicals and the formation of laddar polymers start around 250°C, aromatization and further cyclization occur due to the release of NH radicals associated with the out-gases of NH_3 and HCN, and at 400–700°C aromatization due to dehydrogenation results in the formation of ribbons, which leads to the growth of aromatic molecules due to a denitrogenation reaction, and forms basic structural units (BSU) in carbon materials [Oberlin, 1984, 1989; Guigon et al., 1984a, b]. The structural units after carbonization above 1300°C are stacks of distorted layers with their localized orientation. In order to produce carbon fibers, stabilization of spun PAN fibers is essential, which is usually carried out in air at 200–300°C and its fundamental chemical reactions are considered to be oxidation. Various chemical and physical modifications are expected to occur, such as formation of an intermolecular network connected by oxygen, formation of cyclized structures, polymerization of nitrile side groups to a naphthyridine ring, shrinkage of the fibers and occurrence of shrinkage stress. Detailed reactions remain controversial in a number of published papers.

During stabilization and carbonization, the stress was pointed out to be accumulated mainly due to hindrance of the thermal motion of PAN molecules, shrinkage caused by oxidation reactions during stabilization and also shrinkage by pyrolysis and carbonization accompanied by the departure of hydrocarbon gases. The direct measurement of this stress in PAN fibers has been carried out [Ogawa and Saito, 1995]. A scheme of the changes in stress during heating to 600°C with a rate of 4.5°C min^{-1} in air and in a nitrogen atmosphere is shown in Fig. 4-5.

The changes in stress with temperature may be divided into four regions, as indicated in Fig. 4-5. In region I, stress increases to a maximum (region I-1) and then relaxes down to a minimum at around 240°C (region I-2), which is supposed to be due to the motion of

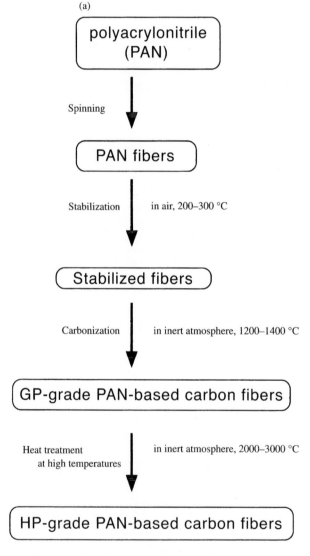

Figure 4-3 Production procedures for carbon fibers: (a) PAN-based carbon fiber.

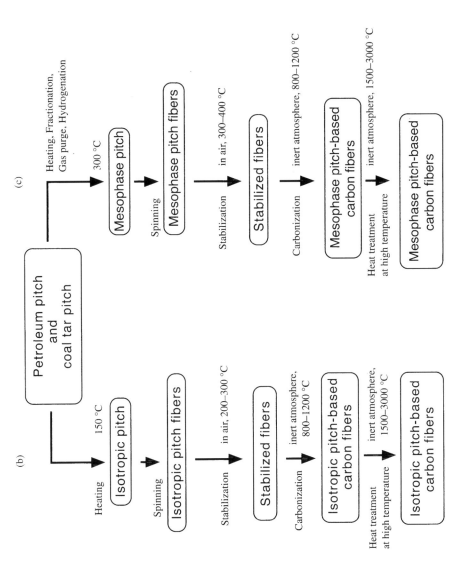

Figure 4-3 (Contd) Production procedures for carbon fibers: (b) isotropic-pitch-based and (c) mesophase-pitch-based carbon fibers.

PAN molecules at high temperatures. If the fibers were cooled from this region, the shrinkage stress always became zero. For example, cooling down from the point A in Fig. 4-5, a little before reaching the minimum, results in zero stress below 180°C. Region II, an abrupt increase in stress, reaching a maximum again, is concluded to be related to the change in chemical structures in the fibers, such as network formation by oxygen, and polymerization of the nitrile side-group. The cooling from this region leaves certain stress in the fibers, which becomes constant below 180°C, as shown by the points B and B', C and C' in Fig. 4-5. In Fig. 4-6a and b, the changes in shrinkage stress with time at a constant temperature in air and in an atmosphere with various oxygen contents at 242°C, respectively, are shown. The scheme of stress change is exactly the same as that with a constant rate heating, but a pronounced dependence on oxidation conditions, temperature and oxygen content, is observed in the second region. From these experimental results, a parameter, stabilization index SI, which is a ratio of the stress at 180°C in cooling (e.g. point B') to that at the temperature where cooling started (point B), was proposed to evaluate the degree of modification in chemical structure in fibers by oxidation (stabilization) [Ogawa, 1994b; Ogawa and Saito, 1995]. This parameter SI is found to have a good correspondence to the content of oxygen after stabilization and also to the tensile strength of the resulting carbon fibers, as shown in Fig. 4-7a and b, respectively.

In region III in Fig. 4-5, the fibers show different behaviors in stress, depending on atmosphere; in air stress decreases quickly and reaches a minimum at a low temperature,

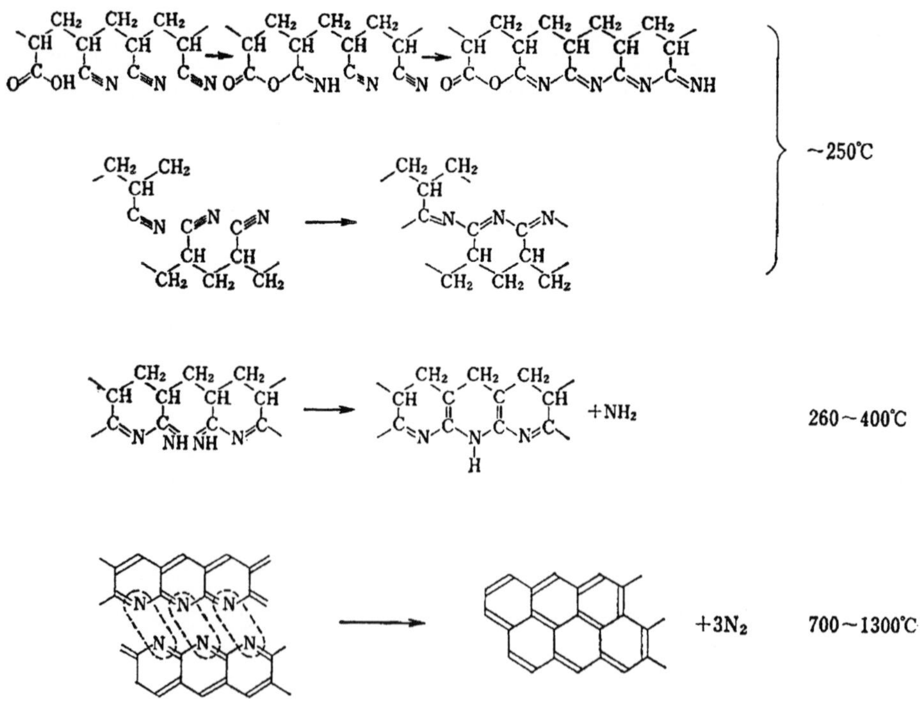

Figure 4-4   Changes in the molecular structure of PAN at high temperatures.

but in nitrogen it decreases to lower minimum at a slightly higher temperature. This region was supposed to correspond to further pyrolysis and plasticization of polymers formed in previous regions. Region IV, with a gradual increase in stress, was supposed to be due to the formation and growth of carbon hexagonal layers.

Because of plasticization during stabilization and carbonization, stretching of fibers during these processes was found to be effective in improving the performance of carbon fibers [Watt and Johnson, 1969], which was understood to be caused by the improvement in preferred orientation of structural units of distorted layer stacks along the fiber axis. Figure 4-8 shows how the improvement in orientation is effective in achieving a high modulus by plotting the modulus of fibers against orientation degree (width of half-maximum of the orientation function) $\Phi_{1/2}$. The effect of stretching during high-temperature heat treatment on the structure and texture of carbon fibers will be discussed later. From the analysis of stress in fibers shown in Fig. 4-5, stabilization in keeping the fiber length constant, where the fibers are under tension in region I-2 and under shrinkage in region II, and carbonization under stretching in the temperature range of region III and IV was found to be effective in achieving a high mechanical strength with a shortened processing time [Ogawa and Saito, 1995].

(b) Pitch-based carbon fibers

Pitch-based carbon fibers have to be divided into two classes, as explained before, based not only on the precursor but also on their structure and properties; isotropic-pitch-based

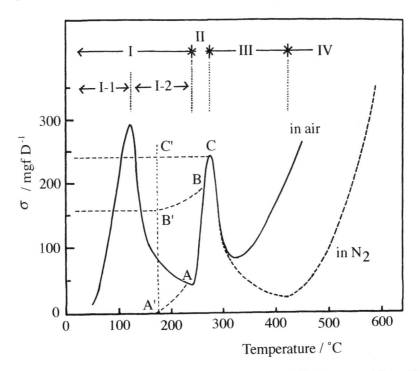

Figure 4-5   Changes in stress σ exerted on the PAN fiber up to 600°C [Ogawa and Saito, 1995].

and anisotropic mesophase-pitch-based carbon fibers. The production process for these two fibers is shown in Fig. 4-3b and c, respectively. Fundamental processes for these carbon fibers are spinning, stabilization and then carbonization, as for PAN-based fibers (Fig. 4-3a).

For the production of isotropic-pitch-based carbon fibers, not so many problems have been pointed out, only the need for a pitch with homogeneity and appropriate fluidity. Some examples of optically isotropic pitches used for carbon fiber production are listed in Table 4-2, showing the softening point, toluene insoluble fraction, percentage of optically anisotropic regions and chemical analysis data [Sato et al., 1993].

In some pitches, the formation of spheres with optically anisotropic texture and their growth in size and number were observed with increasing temperature; these were called mesophase spheres and their detailed texture has been studied [Brooks and Taylor, 1968; Honda et al., 1971; Auguie et al., 1980]. These spheres were known in the next step of heating to coalesce with each other and finally a bulk with anisotropy flow texture was formed, which was called bulk mesophase. In the beginning of the development, mesophase-pitch-based carbon fibers were spun from the bulk mesophase [Fujimaki and Otani, 1976]. In the next step of development, however, various processes, such as fractionation and hydrogen gas purge of pitches, and addition of hydrogen-donating compounds, were introduced in order to improve the spinnability of precursor pitches. The

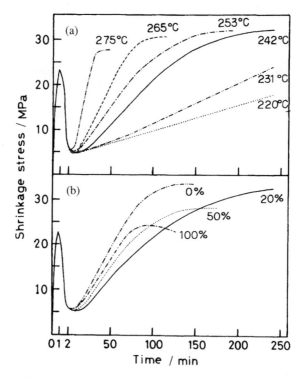

Figure 4-6 Changes in shrinkage stress on the PAN fiber with time at different temperatures (a) and in atmosphere with different oxygen contents (b) [Ogawa and Saito, 1995].

pitches thus prepared were reported to have a performance during heating to high temperatures slightly different from that observed on the pitch without these pretreatments, e.g. in a case being called "dormant pitch" [Otani, 1981]. Therefore, these pitches prepared through hydrogenation might be different in detailed structure and texture from the bulk mesophase, even though all of them are optically anisotropic. However, the carbon fibers prepared from these anisotropic pitches, whatever preparation processes have been employed, have customarily been called "mesophase-pitch-based carbon fibers". Therefore, the same nomenclature was applied in this book. A detailed discussion on the structure and heating performance was presented in a recently published review [Oberlin et al., 1998].

The preparation of optically anisotropic pitches (mesophase pitches) as precursors is the most important process in the production of carbon fibers, as discussed above. In Table 4-2, the characteristics of some mesophase pitches are shown in comparison with isotropic pitches. The factors influencing carbon fiber production are known to be the content of

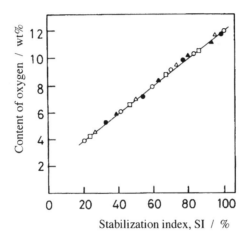

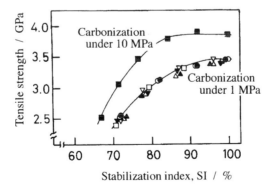

Figure 4-7 Dependence of oxygen content after stabilization (a) and tensile strength of carbon fibers (b) on the stabilization index SI of PAN fibers prepared by using different comonomers such as methyl acrylate, sodium acrylate and acrylamide [Ogawa and Saito, 1995].

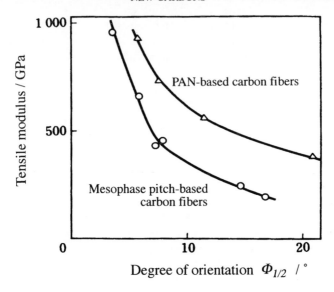

Figure 4-8 Dependence of tensile modulus of carbon fibers on the degree of orientation $\Phi_{1/2}$ of crystallite in the fibers [Inagaki and Hishiyama, 1994].

mesophase (anisotropic ratio, i.e. the percentage of optically anisotropic regions), softening point, degree of hydrogenation and thermal history of the pitch, on the basis of extensive works by various authors. These factors are related to each other, an example being shown by the relation among softening point, mesophase content and hydrogenation treatment in Fig. 4-9. This figure shows that the softening point of the pitch markedly decreases with the decrease in mesophase content and also by hydrogenation treatment. In

Table 4-2. Pitches used as precursors for carbon fiber production [Sato et al., 1993]

Pitches	Isotropic			Anisotropic				
Origin of pitch	Coal	Coal	Petro	Coal	Coal	Coal	Synth	Synth
Softening point (°C)	267	250	250	304	308	299	257	265
Toluene insoluble (wt%)	74.6	44.1	58	81	86	95	43	74
Anisotropic ratio (%)	0	0	–	94	100	100	100	100
Elemental composition								
C	94.3	93.4	93.9	94.9	94.2	95.2	94.4	95.2
H	3.89	4.44	4.65	3.69	3.92	3.82	5.91	5.02
N	1.01	0.85	0.2	0.68	1.05	0.89	< 0.1	0.1
S	0.32	0.29	0.7	0.16	0.28	0.15	< 0.01	0.16
H/C ratio	0.5	0.57	0.59	0.47	0.50	0.48	0.75	0.63

Coal: coal tar pitch; Pitch: petroleum pitch; Synth: synthetic pitch.

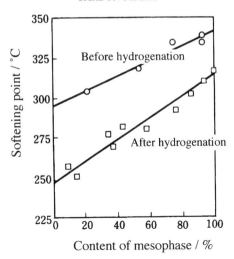

Figure 4-9 Dependence of the softening point of pitch on mesophase content before and after hydrogenation [Sato et al., 1993].

order to be able to spin the fibers at low temperatures, where the thermal stability of the pitch can be assured, hydrogenation is a fundamental process. After hydrogenation, the pitches were heated to start condensation reactions and to remove volatile matters and, as a consequence, mesophase pitches with mesophase content close to 100% were obtained.

One of the advantages of mesophase-pitch-based carbon fibers is that it is possible to achieve a high degree of orientation of carbon hexagonal layers along the fiber axis during spinning and it is not necessary to stretch the fibers during stabilization and carbonization. This fact reveals the importance of the spinning process for achieving high performance in the resultant carbon fibers. Figure 4-10 shows an example of the change in orientation

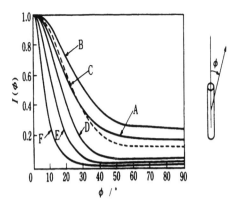

Figure 4-10 Orientation functions measured at different steps of production of mesophase-pitch-based carbon fibers. A: as-spun; B: stabilized; C: 1000°C-treated; D: 1600°C-treated; E: 2000°C-treated; F: 2500°C-treated [Fujimaki and Otani, 1976].

function determined by X-ray diffraction during each step of carbon fiber processing [Fujimaki and Otani, 1976]. Just after spinning a relatively high degree of orientation is already obtained. It is slightly disturbed by stabilization, but improved by heat treatment at high temperatures, the improvement becoming pronounced with the increase in heat treatment temperature.

A further advantage of mesophase-pitch-based carbon fibers is the ability to realize various textures in their cross-sections, as shown schematically in Fig. 4-11. By controlling the viscosity of the pitch, the texture can vary from radial alignment of carbon hexagonal layers with a wedge to an onion-like one (concentric around the fiber axis) through either random or double textures with concentric and radial alignment. In Fig. 4-12, SEM micrographs are shown on the cross-section of the fibers which are produced from exactly the same mesophase pitch by changing the spinning condition. This texture in cross-section has a great influence on the structural change with high-temperature treatment and the properties of fibers, as will be explained in the next section.

By using different shapes of spinneret, pitch-based carbon fibers with non-circular cross-sections have been prepared [Edie et al., 1993, 1994; Wang et al., 1998]. In Fig. 4-13, SEM micrographs of some non-circular carbon fibers, ribbon-like, Y-shape and hollow fibers, are shown. High crystallinity and high thermal conductivity were obtained from carbon fibers with a ribbon-like cross-section [Edie et al., 1994] and better mechanical properties were reported from thin hollow fibers than from solid ones [Wang et al., 1998].

(c) Vapor-grown carbon fibers

Vapor-grown carbon fibers are characterized by the presence of minute iron particles at the top of the fibers, a typical TEM micrograph being shown in Fig. 4-14. The growth conditions and mechanism have been explored by various authors [Oberlin et al., 1976; Endo, 1978; Katsuki et al., 1981; Tibbetes, 1985; Ishioka et al., 1992a] after the pioneering work of Koyama and Endo [1973]. The preparation of vapor-grown carbon fibers has been performed by two methods, seeding catalyst and floating catalyst methods, principles of

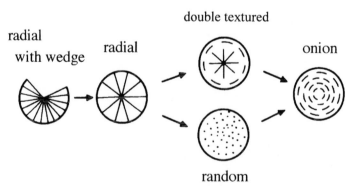

Figure 4-11   Various textures in cross-section of mesophase-pitch-based carbon fibers..

which are schematically illustrated in Fig. 4-15a and b, respectively [Ishioka et al., 1995]. In most cases reported, high-purity hydrogen gas was used to reduce and activate the catalyst metal particles, iron particles in most cases, to synthesize carbon fibers with high yield. The impurity of inert gases, such as nitrogen and argon, and oxidative gases, such as oxygen and water vapor, was found to contaminate the catalyst particles and markedly decrease the yield of fibers. In pure inert gas, no formation of fibers was confirmed.

In the seeding catalyst method, catalyst particles are seeded on a substrate, such as graphite or ceramics, either by painting very fine metal particles which may be prepared by a vacuum evaporation method and suspended in an alcohol, spraying the alcohol solution of metal nitrate and then decomposing at high temperatures, or by forming fine iron particles by the thermal decomposition of organic iron compounds such as ferrocene. By the development of these seeding techniques, the synthesis of vapor-grown carbon fibers becomes more reproducible with a higher yield than the procedure employed by the pioneers [Endo, 1978], where the supply of catalyst iron occurred only by chance.

The metal particles with a size less than 20 nm were found to be useful for growing fibers, with the finer particles giving the higher yields and the faster growth in fiber length. In general, the growth rate in fiber length is roughly 1 mm min^{-1} and that in fiber diameter 5–10 μm h^{-1}. By this seeding catalyst method, fibers a few tens of millimeters in diameter and a few tens of centimeters in length were obtained, but the following disadvantages

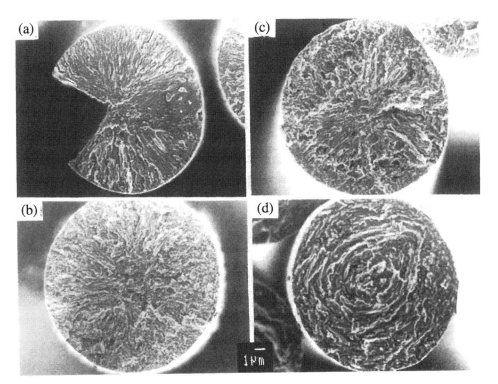

Figure 4-12 SEM micrographs of cross-sections of carbon fibers prepared from the same precursor mesophase pitch [Inagaki et al., 1991a]

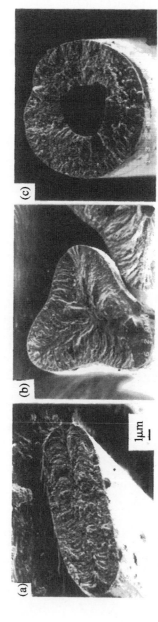

Figure 4-13 SEM micrographs of mesophase-pitch-based carbon fibers with non- circular cross-sections: (a) ribbon-like, (b) Y-shape and (c) hollow carbon fibers. [Courtesy of Prof. Wang of Tianjin University.]

were pointed out: inhomogeneity in the radius and length of fibers, relatively low fiber yield and discontinuity in unit operations for synthesis (i.e. seeding of catalyst, reduction and purification of catalyst metal, growth of fibers and recovery of fibers).

The floating catalyst method was developed in order to improve the fiber yield. In this method, two ways to feed the catalyst particles in the reaction space were employed, direct introduction of fine metal particles and introduction of organometallic compounds which form fine metal particles by thermal decomposition in the reaction space. By using ferrocene in the benzene–hydrogen system, a very high yield of carbon fibers with a diameter of 0.1–1.5 μm and a length of about 1 mm was reported [Endo and Shikata, 1985]. Figure 4-16 shows the dependence of fiber yield on the ferrocene fraction, indicating a suitable amount of ferrocene needed to give a maximum yield [Egashira et al., 1983].

Instead of high-purity hydrogen gas, which is costly, Linz-Donawitz converter gas (LDG), which is an industrial byproduct and consists mainly of CO, was shown to be used successfully as a carrier gas for benzene to synthesize fibers by the floating catalyst

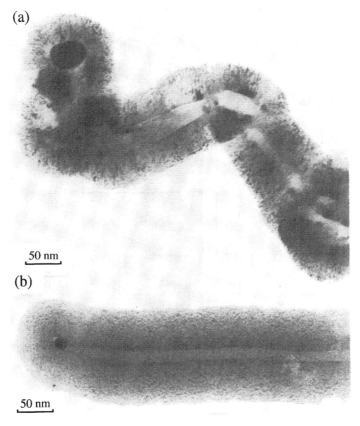

Figure 4-14 TEM micrographs of vapor-grown carbon fibers. [Courtesy of Prof. Endo of Shinshu University.]

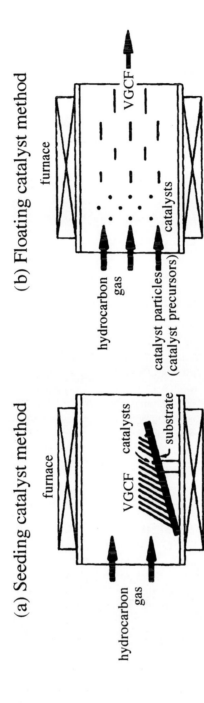

(a) Seeding catalyst method

(b) Floating catalyst method

Figure 4-15 Illustrations of systems for the preparation of vapor-grown carbon fibers by seeding and floating catalyst methods [Ishioka et al., 1995].

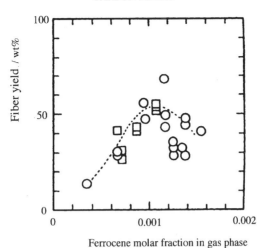

Figure 4-16 Dependence of fiber yield on molar fraction of ferrocene in the catalyst precursor [Egashira et al., 1983].

method [Ishioka et al., 1992c]. The elimination of oxygen and water vapor from LDG (67% CO, 1.2% H_2, 16% CO_2, 15% N_2, 0.2% O_2 and 0.6% H_2O) was essential to obtain fibers with a high yield, about 50 wt% yield of the fibers with a diameter of about 3 μm and length of about 3 mm. Using gas mixtures, the effect of the composition of this carrier gas, such as content of CO and CO_2, was studied in detail [Ishioka et al., 1992a, b]. When a mixture of ferrocene, thiophene and cobaltacetylacetonate as a precursor of catalyst particles and a mixed gas of 77% CO, 19% CO_2 and 4% H_2 were used, a high fiber yield of 70 wt% was obtained. In particular, the content and the metal species of metalacetylacetonate are found to govern the yield, as shown in Fig. 4-17.

The vapor-grown carbon fibers prepared by the floating catalyst method were usually thin but homogeneous in diameter and also short in length, in comparison with those by the seeding catalyst method. By controlling the gas flow pattern in the furnace and temperature of the inlet gas, straight fibers could be synthesized, although most of the fibers synthesized by the floating catalyst method were crooked [Ishioka et al., 1993a, b]. SEM micrographs of the fibers with two morphologies are shown in Fig. 4-18. From the industrial point of view, the floating catalyst method seems to be promising because of the continuous operation of the synthesis, high fiber yield and, as a consequence, low cost.

By the catalytic decomposition of different precursor gases, not only hydrocarbons but also carbon monoxide, on minute metal particles, carbon fibers were reported to be formed [Rodriguez, 1993], which have sometimes been called carbon filaments and recently even carbon nanofibers and/or carbon nanotubes. Their diameter, appearance and texture depend strongly on the decomposition conditions, decomposition temperature, catalyst metals, precursor gases, etc. [Baker et al, 1975; Audier et al., 1981; Motojima et al., 1989a, b]. In Fig. 4-19, SEM micrographs of carbon helical microcoils are shown, which have been prepared by the catalytic pyrolysis of acetylene in the presence of a metal powder catalyst and a small amount of phosphorus impurity [Motojima et al., 1995]. Their

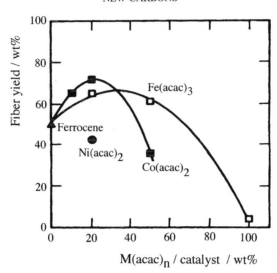

Figure 4-17   Dependence of fiber yield on the content of various metalacetylacetonate M(acac)$_n$ in the catalyst precursor [Ishioka et al., 1992b].

possible applications are being extensively studied [Motojima et al., 1998]. Recently, carbon filaments with different textures were synthesized by the catalytic decomposition of CO on metal at 550°C [Soneda and Makino, 1998]. By changing the ratio of CO to H$_2$, different textures of carbon layers, from parallel to perpendicular to the filament axis, were realized. In Fig. 4-20, an example of a filament where carbon layers are stacked perpendicular to the axis is shown.

4.3. Structure

(a) PAN-based carbon fibers

Using high-resolution TEM techniques, the structure of commercial PAN-based carbon fibers of HM and HT types was studied in detail [Oberlin et al., 1988]. Representative *002*

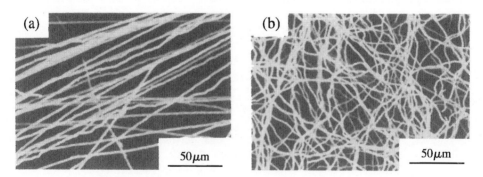

Figure 4-18   SEM micrographs of vapor-grown carbon fibers with two morphologies, straightened (a) and crooked (b) [Ishioka et al., 1993a].

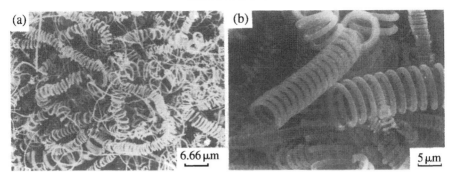

Figure 4-19   SEM micrographs of carbon microcoils by the pyrolysis of acetylene in the presence of Ni (a) and Mo (b), with a small amount of PCl$_3$ at 650°C. [Courtesy of Prof. Motojima of Gifu University.]

lattice fringes and *002* dark-field images of these two types of fiber are shown in Figs 4-21 and 4-22. It is clearly shown that there is a pronounced difference in structural units of stacked layers between HM and HT types; in the former the units are thicker and longer than those in the latter. In addition, the curvature of the fringes much larger in the former. This difference is supposed to be caused by the difference in thermal history between these two types of fibers; the former has experienced high temperatures, possibly under stretching, but the latter relatively low temperatures [Guigon et al., 1984a, b]. However, the exact reason is not known because detailed preparation conditions on these commercial fibers have not been published.

From a detailed analysis over a whole cross-section of the fibers, the structure models for HM and HT types of fibers were constructed [Guigon et al., 1984a, b]. The proposed models are reproduced in Fig. 4-23. In HM type, carbon layers are entangled with each other in the central part of the fiber, but they are straighter and stack parallel in a wider range in the periphery (Fig. 4-23a). In HT type, the carbon layers are much smaller and thinner, with no marked difference between central and peripheral parts (Fig. 4-23b).

The effect of stretching during heat treatment at high temperatures is exhibited through the measurements of magnetoresistance, as shown in Table 4-3 [Hishiyama et al., 1984]. PAN-based carbon fibers show only negative $(\Delta\rho/\rho)_{cr}$ after heat treatment to 3000°C, but it becomes positive even after 2700°C treatment on the fibers stabilized and carbonized under a load of 40 mg per filament. It is clear that this marked increase in $(\Delta\rho/\rho)_{cr}$ is due to the development of layers along the fiber axis [Hishiyama et al., 1991]. It has to be pointed out here that the anisotropy ratio r_T, which is a measure of orientation of the crystallite perpendicular to the fiber axis, decreases in value by stretching, although the anisotropy ratio r_{TL}, a measure of orientation along the fiber axis, is kept small. This suggests that the orientation scheme of carbon layers in the fiber changes from an axial to a planar orientation by stretching during heat treatment processes. This expectation was proved by SEM observation on one of the fibers shown in Table 4-3, where the cross-section became rectangular and ribbon-like, and graphite layers were seen to orient parallel to the long edge of the rectangle [Hishiyama et al., 1984].

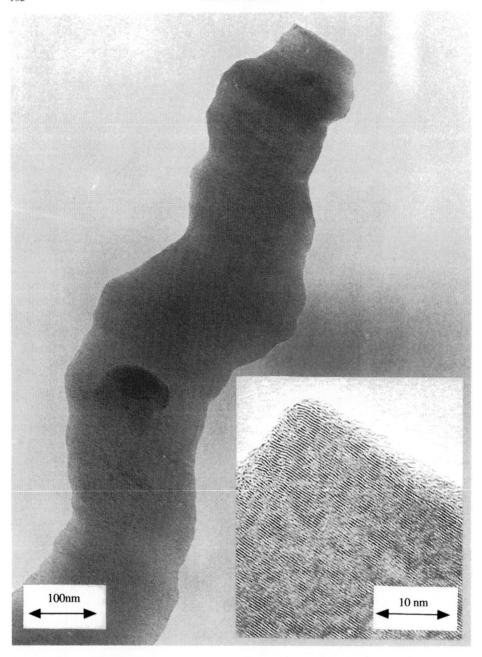

Figure 4-20   TEM micrographs of the filament prepared from a mixture of 80% CO and 20% H_2 at 550°C by catalytic decomposition. [Courtesy of Dr Soneda of National Institute for Resources and Environment.]

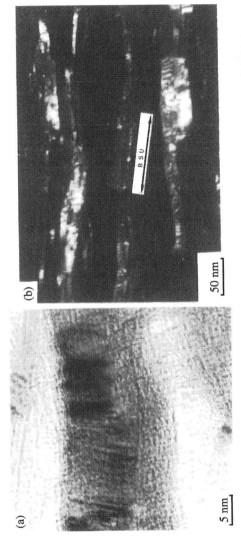

Figure 4-21 TEM micrographs of HM type of PAN-based carbon fibers: (a) *002* lattice fringes, (b) *002* dark-field images. [Courtesy of Mme Oberlin.]

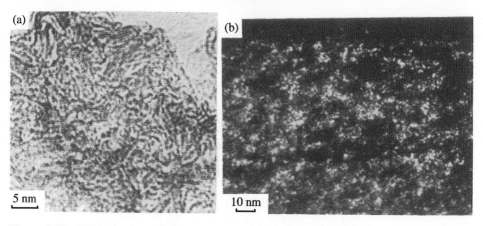

Figure 4-22   TEM micrographs of HT type of PAN-based carbon fibers: (a) *002* lattice fringes, (b) *002* dark-field images. [Courtesy of Mme Oberlin.]

(b) Pitch-based carbon fibers

Because of the optically isotropic precursor pitch, the isotropic-pitch-based carbon fibers have an isotropic texture in their cross-section after carbonization and even after heat treatment at high temperatures, as shown in Fig. 4-24.

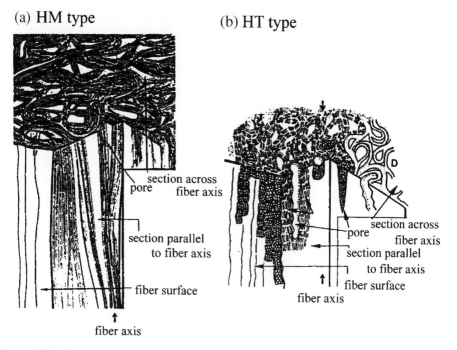

Figure 4-23   Structure models for HM (a) and HT (b) types of PAN-based carbon fiber. [Courtesy of Mme Oberlin.]

Table 4-3. Effect of stretching on magnetoresistance parameters of PAN-based carbon fibers [Hishiyama et al., 1984]

Stretching conditions	HTT(°C)	$(\Delta\rho/\rho)_{cr}(\%)$	r_{TL}	r_T
Without stretching	3000	− 0.547	0.063	0.977
	2900	− 0.954	0.049	0.994
	2800	− 0.618	0.056	0.864
40 mg/filament during stabilization and carbonization (PAN-S)	3000	26.18	0.027	0.753
	2830	6.37	0.022	0.403
40 mg/filament during 2700°C treatment of PAN-S (PAN-SG-4)	3000	26.20	0.110	0.200
	2830	9.19	0.011	0.303
	2700	2.10	0.045	0.545
80 mg/filament during 2700°C treatment of PAN-S (PAN-SG-8)	3000	21.18	0.225	0.665
	2830	9.23	0.023	0.945
	2700	4.74	0.000	0.789
120 mg/filament during 2700°C treatment of PAN-S (PAN-SG-12)	3000	22.99	0.083	0.874
	2830	4.73	0.091	0.209
	2700	1.54	0.000	0.540

The isotropic-pitch-based carbon fibers are used in large and important applications as activated carbon fibers with a very high surface area, as will be explained in the next chapter. The isotropic texture of fibers is advantageous for activation and the fibrous morphology gives effective contact with adsorbates, either gas or liquid.

In contrast, the mesophase-pitch-based carbon fibers are optically anisotropic in their cross-section. As mentioned before, it is one of their characteristics to be able to have different textures from the random alignment of carbon layers to concentric and radial ones in their cross-sections. In Fig. 4-25, the texture of a cross-section is shown by using SEM on the fibers prepared from the same mesophase pitch with different spinning conditions and heat-treated up to 3000°C, of which cross-sectional textures after carbonization have been shown in Fig. 4-12 [Inagaki et al., 1991a]. Radial and concentric textures are seen more clearly, the radial texture with a wedge showing distinct straightness of the layers. On the fiber with radial texture, the wedge shows a larger

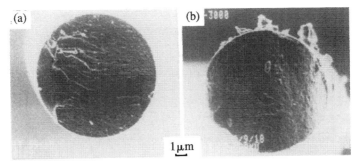

Figure 4-24 SEM micrographs of a fractured cross-section of isotropic-pitch-based carbon fiber: (a) 2600°C-treated and (b) 3000°C-treated. [Courtesy of Prof. Y. Hishiyama and A. Yoshida of Musashi Institute of Technology.]

opening after high-temperature treatment (Fig. 4-25a) than in the carbonized one (Fig. 4-12a). This difference in texture reflects the development of a graphitic structure with 3000°C treatment, and various parameters as a measure of degree of graphitization are summarized in Table 4-4. The fibers with the radial texture with a wedge (straight radial, I-1-30) have high values of graphitization degree P_1, crystallite sizes along the a- and c-axes, L_a and L_c, and also a positive value of magnetoresistance $(\Delta\rho/\rho)_{cr}$, whereas the fibers without a wedge, i.e. zigzag radial (I-2-30), have small values of these parameters.

Figure 4-25 SEM micrographs of cross-sections of carbon fibers prepared from the same precursor mesophase pitch after heat treatment at 3000°C [Inagaki et al., 1991a].

Table 4-4. Structural parameters of mesophase-pitch-based carbon fibers prepared from the same precursor mesophase pitch with different cross-sectional textures after heat treatment at 3000°C (refer to Fig. 4-25) [Inagaki et al., 1991a]

Carbon fiber	d_{002} (nm)	$L_c(002)$ (nm)	$L_a(110)$ (nm)	P_1	P_{ABA}	P_{ABC}	$(\Delta\rho/\rho)_{cr}$ (%)	r_T	r_{TL}
I-1-30	0.3367	49	81	0.41	0.25	0.04	12.447	0.681	0.01
I-2-30	0.3385	24	40	0.30	0.10	0.03	0.372	0.297	–
I-3-30	0.3380	27	45	0.32	0.12	0.01	0.849	0.597	–
I-4-30	0.3379	27	56	0.35	0.13	0.04	0.352	0.480	–

From high-resolution TEM analysis, it was shown that texture in the nanometric scale in the precursor mesophase pitch was retained in the fibers after spinning and carbonization, even after 3000°C treatment [Oberlin and Guigon, 1988; Inagaki et al., 1991a]. In 1200°C-carbonized fibers, two regions with different nanotextures, with small layers being oriented in one region and random in another, are observed, as shown on the fibers with radial-with-wedge texture in Fig. 4-26. These two regions are reasonably supposed to come from anisotropic and isotropic parts, respectively, in the precursor mesophase pitch. By heat treatment at high temperatures, it was shown from TEM micrographs that the layers grow in the oriented region, but in the random region the growth of layers is so restricted that it becomes porous.

These experimental results by different techniques reveal that the cross-sectional texture, which is determined by the texture of precursor mesophase pitch and the conditions during its spinning, is retained even after heat treatment at temperatures as high as 3000°C and governs the structural development with heat treatment. In other words, the preparation of the precursor pitch and its spinning are decisive unit operations to control the structure and properties of mesophase-pitch-based carbon fibers.

(c) Vapor-grown carbon fibers

Characteristics in the structure of vapor-grown carbon fibers are summarized into three points: concentric alignment of carbon hexagonal layers along the fiber axis, the presence

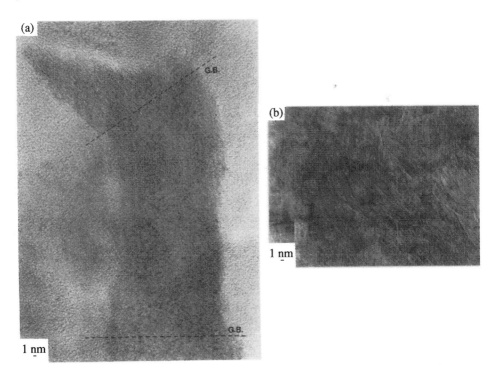

Figure 4-26 TEM micrograph of mesophase-pitch-based carbon fibers. [Courtesy of Mme Oberlin.]

of a catalyst particle at the tip of fibers and the formation of a hollow tube at the center
of the fibers, as can be seen in the micrograph of Fig. 4-27. The diameter of the hollow
tube is very similar to that of the catalyst particles located at the tip of the fiber, but always
slightly smaller. As shown in Fig. 4-27, detailed high-resolution electron microscopic
studies reveal that the wall of the hollow tube consists of long and straight layers with
turbostratic stacking and the outer part is a piling-up of small BSU with a size of about

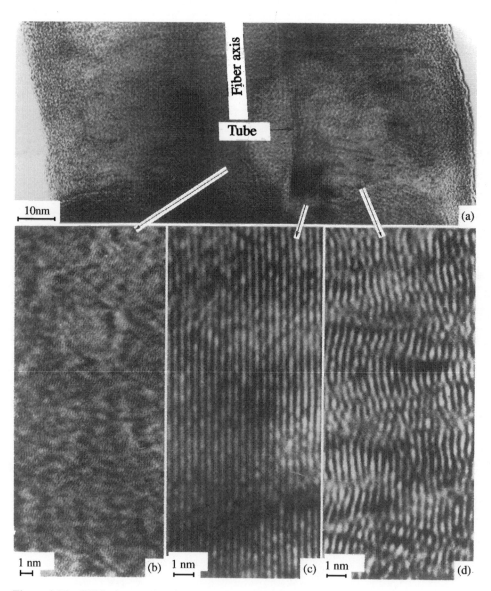

Figure 4-27   TEM micrographs of vapor-grown carbon fibers. [Courtesy of Prof. Endo of Shinshu
University.]

1 nm and parallel stacking of two to three carbon layers [Oberlin et al., 1976]. The structure of the outer part is exactly the same as that of pyrolytic carbons deposited on a substrate through thermal pyrolysis of hydrocarbon gases. These structural characteristics were confirmed not to depend on the synthesis method, hydrocarbon gas or catalyst metal used.

From these structural studies, the growth model shown in Fig. 4-28a has been proposed for the seeding method [Oberlin et al., 1976], which can be expanded to that for the floating catalyst method with a slight modification (Fig. 4-28b). First, either metal oxide particles are reduced to metal or the surface of metal particles is cleaned up by hydrogen gas. In this step, the metal particles formed are in the liquid phase and either stay on the

(a) Seeding catalyst method

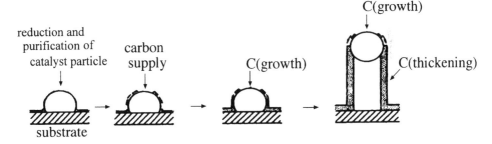

(b) Floating catalyst method

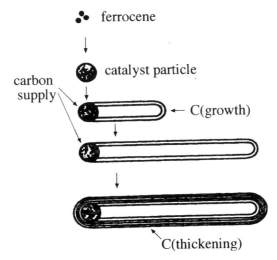

Figure 4-28   Models for the growth of vapor-grown carbon fibers by the seeding and floating catalyst methods [Ishioka et al., 1995].

substrate or are floating. Carbonaceous materials formed by the pyrolysis of hydrocarbon gas are supposed to be supplied onto these liquid metal particles. Carbonaceous materials may diffuse along the surface of metal particles and at the same time pyrolysis proceeds by some catalytic effect of metal to form hexagonal carbon layers. In the seeding catalyst method, the carbon layers thus formed are deposited at the boundary between the substrate and catalyst particle. As a consequence of the continuous supply of carbonaceous materials, catalyst particles are raised from the substrate and thin fibers (tubes) are formed. The formation of these thin tubes seems to be terminated if the surface of the metal particle is covered by deposited carbon or if it is contaminated by an impurity such as oxygen. In the floating catalyst method, the carbon layers formed on the surface of catalyst particles are deposited at the boundary between the already accumulated carbon and the catalyst particle. Therefore, it results in a fibrous texture. In the next step, pyrolytic carbon is deposited onto the surface of these thin fibers, and the fibers grow in the radial direction. Therefore, the growth of these fibers occurs in two steps, i.e. growth along the fiber length and then along the radius. It was shown experimentally that the former step occurs in a short period but the latter takes time [Endo et al., 1976; Tibbetts et al., 1986].

The formation of thin tubes at the beginning of the growth of vapor-grown carbon fibers was proved experimentally from the observation of their pull-out on broken surfaces of the fibers [Endo et al., 1995]. These thin tubes are now understood to be carbon nanotubes which are formed by a catalytic method. Since the first report on the formation of nanotubes by a discharging arc between carbon electrodes [Iijima, 1991], much attention has been paid to their synthesis and properties, and various synthesis processes have been proposed. One recent interesting result was the formation of bundles with oriented nanotubes from a single crystal of SiC under high vacuum [Kusunoki et al., 1997].

Another feature of the vapor-grown carbon fibers is high graphitizability, the marked development of a graphitic structure with heat treatment. In Fig. 4-29, interlayer spacing d_{002} is plotted against heat treatment temperature (HTT) on vapor-grown carbon fibers prepared from benzene by the seeding catalyst method, anthracene coke, which is a typical graphitizing carbon, and PAN-based carbon fibers [Endo et al., 1976]. Very high graphitizability of the vapor-grown carbon fibers is clearly seen, even better than for anthracene coke at high temperatures and markedly different to PAN-based carbon fibers.

The graphitizability is also known to depend on the diameter of the fibers. In Fig. 4-30, changes in the graphitization degree P_1 with HTT are shown on the vapor-grown carbon fibers with different diameters [Iwashita et al., 1993]. The fibers with diameters less than 1 μm show a rather low degree of graphitization. With the heat treatment, the morphology of the fibers also changes. In Fig. 4-31, SEM micrographs of cross-sections of the fibers are shown as a function of HTT. Around 3000°C, the fibers are polygonized, which is due to the growth of graphite layers [Yoshida et al., 1995].

4.4. Properties and Applications

(a) Mechanical properties

Mechanical properties are the most important for carbon fibers, as they are used for the classification of carbon fibers (Fig. 4-1). The industrial effort in the production of carbon

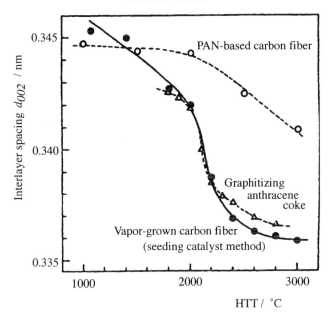

Figure 4-29 Dependence of interlayer spacing d_{002} on HTT for vapor-grown and PAN-based carbon fibers compared with a graphitizing anthracene coke. [Courtesy of Prof. Endo of Shinshu University.]

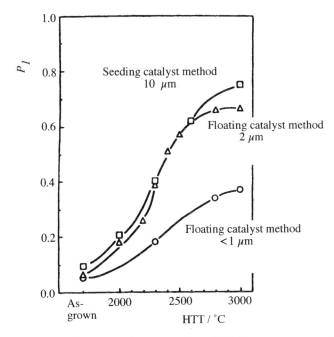

Figure 4-30 Dependence of graphitization degree P_1 on HTT for vapor-grown carbon fibers with different diameters [Iwashita et al., 1993].

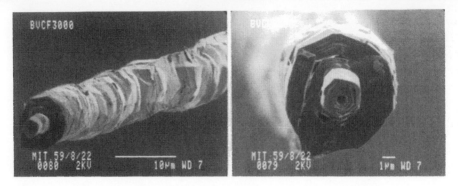

Figure 4-31 SEM micrograph of vapor-grown carbon fiber after high-temperature treatment.
[Courtesy of Prof. Hishiyama and Dr Yoshida of Musashi Institute of Technology.]

fibers in various respects resulted in a great improvement in mechanical properties and
also in their reproducibility. Figure 4-32 shows the changes in tensile strength on
commercial PAN-based carbon fibers produced in a company; since 1985 different grades
with improved strength have been commercialized and the tensile strength has reached
more than 7 GPa, about double that before 1985.

It is well known that these mechanical properties of carbon fibers depend strongly not
only on the precursor but also on heat treatment conditions, mainly its temperature, as do

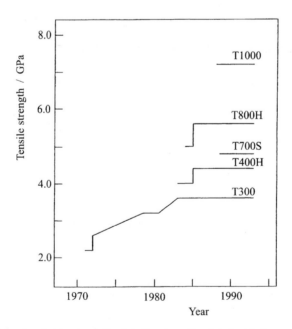

Figure 4-32 Yearly development of the tensile strength of industrially produced PAN-based
carbon fibers. [Courtesy of Toray Co. Ltd.]

Table 4-5. Characteristics of four kinds of carbon fibers

	Isotropic-pitch-based		PAN-based		Mesophase-pitch-based	Vapor-grown	
	C	G	C	G	G	C	G
Diameter (μm)	14.5	14.5	7.0	6.5	10	5-8	5-8
Density (g cm^{-3})	1.65	1.57	1.74	1.81	2.0	1.80	2.02
Tensile strength (MPa)	720	600	3300	2500	2100	3000	5000 −8000
Tensile modulus (GPa)	32	30	230	300	520	300	500
ρ ($\times$ 10^{-3} Ωcm)	15	5	1.6	0.8	0.5	1	0.05
CTE ($\times$ 10/°C)	1.7	1.5	− 0.7	− 1.2	− 1.2	–	–

C: carbonized; G: high-temperature-treated; ρ: electrical resistivity;
CTE: thermal expansion coefficient.

other properties. For most commercially available carbon fibers, however, these conditions have never declared. Therefore, a detailed understanding of the mechanical properties as a function of heat treatment conditions is limited to data obtained in laboratories.

In Table 4-5, engineering properties of carbon fibers are summarized as functions of the precursor, isotropic pitch, PAN, mesophase pitch and vapor-grown, and also either carbonized (heat-treated to approx. 1300°C) or graphitized (heat-treated to more than 2800°C, irrespective of whether a graphitic structure is developed or not). In each precursor, the density increases and tensile strength and electrical resistivity decrease by graphitization treatment (i.e. high-temperature treatment). However, the precursor predominantly governs these properties; isotropic-pitch-based carbon fibers have a lower density, lower strength and higher resistivity than vapor-grown ones.

Cross-sections of mesophase-pitch-based carbon fibers prepared from the same precursor mesophase pitch by different conditions of spinning are shown in Figs 4-12 and 4-25 [Inagaki et al., 1991a]. On these carbon fibers, mechanical properties and thermal expansion coefficients were measured [Tanabe et al., 1991]. Although the cross-sectional texture of the fibers is different, mechanical and thermal properties were not very different; Young's modulus after high-temperature treatment increased markedly in the fiber of radial texture with a wedge. Young's modulus of these fibers was calculated, employing a simple model by taking account of the orientation function of crystallites along the fiber axis. The calculated modulus shows a good correspondence to the measured one, as shown in Fig. 4-33, indicating that the modulus of carbon fibers is predominantly governed by the preferred orientation of crystallites and also their sizes, which have been pointed out by various authors.

(b) Electromagnetic properties

The change in structure of carbon fibers depends strongly on the precursors, as described in the previous section. For four kinds of carbon fibers, magnetoresistance parameters are summarized in Table 4-6 as a function of HTT. Isotropic-pitch-based carbon fibers show

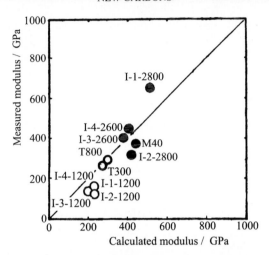

Figure 4-33   Correspondence between tensile modulus measured and calculated by taking account of crystallite orientation in mesophase-pitch-based carbon fibers [Tanabe et al., 1991].

negative magnetoresistance and r_{TL} value of around 0.5, although r_T is close to 1 even after the heat treatment up to 3000°C, indicating a turbostratic structure with a small crystallite size and a lower degree of axial orientation. PAN-based carbon fibers also have negative values of $(\Delta\rho/\rho)_{cr}$ even after 3000°C treatment, but axial orientation is formed. Stretching during spinning of PAN fibers and their heat treatment is clearly shown in Table 4-3 to be

Table 4-6. Magnetoresistance parameters measured on carbon fibers prepared from different precursors by heat treatment at different HTT [Hishiyama et al., 1984]

Carbon fibers	HTT (°C)	$(\Delta\rho/\rho)_{cr}$ (%)	r_{TL}	r_T
Isotropic-pitch-based (IP)	3000	− 0.230	0.615	0.988
	2800	− 0.152	0.512	0.992
	2600	− 0.075	0.679	0.982
PAN-based (PAN)	3000	− 0.547	0.063	0.977
	2900	− 0.954	0.049	0.994
	2800	− 0.618	0.056	0.964
Mesophase-pitch-based (MP)	3000	1.61	0.012	0.417
	2800	− 0.763	0.079	0.929
	2500	− 0.469	0.050	0.858
	2000	− 0.214	0.119	1.000
Vapor-grown (VG)	3000	153.0	0.152	0.854
	2800	138.7	0.047	0.883
	2600	38.6	0.026	0.843
	2400	11.7	0.069	0.952
	2200	0.963	0.000	0.915
	2000	− 1.20	0.087	0.913
	1800	− 0.275	0.000	0.897

very effective in improving the structure, even in PAN-based carbon fibers, as discussed in the previous section. In mesophase-pitch-based and vapor-grown carbon fibers, a transition from a negative to positive value of $(\Delta\rho/\rho)_{cr}$ is observed with increasing HTT; mesophase-pitch-based carbon fibers need much higher HTT in order to reach positive values than vapor-grown ones. For vapor-grown carbon fibers with HTT of 3000°C, slight deviation from the axial orientation of crystallites is observed, which is reasonably supposed to be due to the polygonization of the cross-section of fibers, as shown in Fig. 4-31.

A general scheme of the change in $(\Delta\rho/\rho)_{cr}$, i.e. growth and improvement in the stacking order of carbon layers, was obtained. In Fig. 4-34, the results are shown by plotting $(\Delta\rho/\rho)_{cr}$ against HTT reduced by referring to that for vapor-grown carbon fibers [Hishiyama et al., 1984]; the data in Tables 4-3 and 4-6 are included in this figure. From the shift along the HTT axis, one can determine the heat treatment temperature required in order to obtain the same $(\Delta\rho/\rho)_{cr}$-value, i.e. the same structure. For example, the structure in 3000°C-treated PAN-based carbon fibers corresponds to that obtained by the heat treatment of vapor-grown carbon fibers only at 2200°C. Even by heat treatment under stretching, the increase in $(\Delta\rho/\rho)_{cr}$, i.e. the development in graphitic structure, in PAN-based carbon fibers is restricted, corresponding to that in 2500°C-treated vapor-grown carbon fibers. The structure development in isotropic-pitch-based carbon fibers is strongly depressed, the structure in 1800°C-treated vapor-grown carbon fibers being obtained after their heat treatment at 3000°C. Since mesophase-pitch-based carbon fibers have much better orientation along the fiber axis without stretching, they could have $(\Delta\rho/\rho)_{cr}$ values similar to those obtained for PAN-based ones after stretching, but correspond to only 2200°C-treated vapor-grown ones.

As shown in Fig. 4-34, vapor-grown carbon fibers can cover the range of magnetoresistance from negative to high positive values. This is the case for all electromagnetic properties of these fibers, not only for magnetoresistance (Fig. 4-34) but also for other electromagnetic properties. In Fig. 4-35, electrical resistivity ρ, which is measured on a single filament at room temperature along the fiber axis, is plotted against HTT and compared with carbon prepared from graphitizing coke [Koyama and Endo, 1974]. The change in ρ with HTT observed on these fibers is very similar to a graphitizing coke, indicating their high graphitizability, which corresponds to structural change in these fibers as shown in Fig. 4-29, and the lower value is due to a high degree of orientation of the crystallites along the fiber axis.

For the vapor-grown carbon fibers heat-treated at different temperatures, Fig. 4-36 shows magnetic field dependence of $(\Delta\rho/\rho)_{max}$ at liquid nitrogen temperature and Fig. 4-37 shows the changes in magnetoresistance $(\Delta\rho/\rho)_{max}$ with *TL*- and *T*-rotation under a magnetic field of 1 T at liquid nitrogen temperature [Hishiyama et al., 1984]. With increasing HTT, $(\Delta\rho/\rho)_{max}$ increases its absolute value with a negative sign, and above 2200°C changes its sign to positive and increases markedly; their $(\Delta\rho/\rho)_{cr}$ are plotted in comparison with other carbon fibers in Fig. 4-34. The preferred orientation of the crystallites along the fiber axis becomes marked with the increase in HTT, but its axial orientation scheme is kept in the whole range of HTT, with a large change with *TL*-rotation but almost constant with *T*-rotation. Above 2800°C, however, the change in $\Delta\rho/\rho$

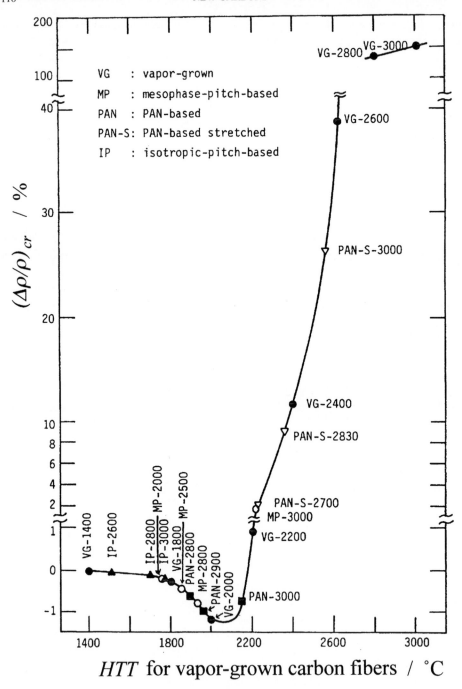

Figure 4-34   Change in the crystallite magnetoresistance $(\Delta\rho/\rho)_{cr}$ of various carbon fibers heat-treated at different temperatures against HTT for vapor-grown carbon fibers [Hishiyama et al., 1984]. The sample codes used are referred to in Table 4-6.

with T-rotation becomes irregular, which is reasonably supposed to come from polygonization of this fiber, as shown in Fig. 4-31.

The electromagnetic properties also depend on diameter of the fiber, as well as HTT and residence time. Figure 4-38 shows the dependence of $(\Delta\rho/\rho)_{max}$ on the diameter of vapor-grown carbon fibers at different HTT and residence times. The larger diameter gives the higher $(\Delta\rho/\rho)_{max}$, i.e. the more marked development of the graphitic structure.

One characteristic of mesophase-pitch-based carbon fibers is their high thermal conductivity, the value of $100 \ \Omega m^{-1} \ K^{-1}$ at room temperature being reported. Its temperature dependence was found to have a good relation with that of electrical resistivity. On a large number of fibers, including commercial and experimental ones, a close relation between thermal conductivity and electrical resistivity is found (Fig. 4-39) [Lavin et al., 1993]. This high thermal conductivity of mesophase-pitch-based carbon fibers can reach almost three times that of metallic copper and silver, and so new applications as heat sinks for electronic devices are expected.

(c) Novel applications to environmental and energy problems

Industrial applications of carbon fibers are summarized in Table 4-7. The main applications of carbon fibers are in various composites, not only with plastics, but also

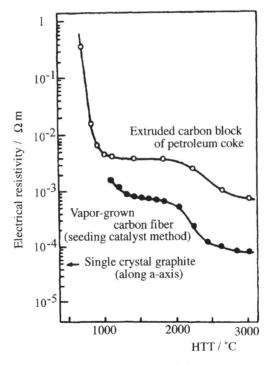

Figure 4-35 Electrical resistivity ρ plotted against HTT for vapor-grown carbon fibers derived from benzene by the seeding catalyst method and extruded carbon block of graphitizing coke [Koyama and Endo, 1974].

with ceramics, including cement, and with carbon. These composites will be described in Chapter 7. Novel and important applications of carbon fibers are shown here, which have been developed recently and studied extensively.

It was found that a large number of microorganisms clung to carbon fibers in a short time [Kojima et al., 1997]. When PAN-based carbon fibers in strands were hung in a solution containing revitalized dirt with 5% drainage (BOD 1000 ppm), microorganisms in the dirt clung quickly to the carbon fibers, forming a large ball as shown in Fig. 4-40 and only a small amount of precipitation. The weight increased roughly 100 times after one day. When other reference fibers, such as cotton, Nylon and polyethylene, were used, no such large amount of clinging was observed.

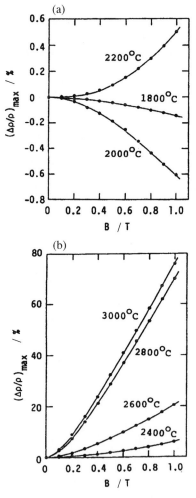

Figure 4-36 Dependence of $(\Delta\rho/\rho)_{max}$ at liquid nitrogen temperature on magnetic field strength B for vapor-grown carbon fibers prepared by a seeding catalyst method [Hishiyama et al., 1984].

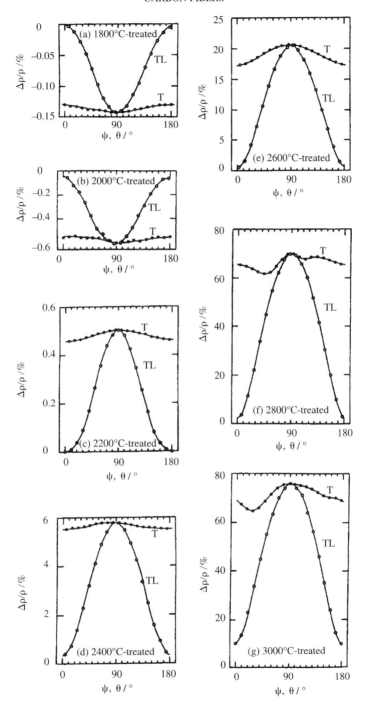

Figure 4-37 Changes in magnetoresistance $(\Delta\rho/\rho)_{max}$ of vapor-grown carbon fibers with *TL*- and *T*-rotation at liquid nitrogen temperature under a magnetic field strength of 1 T [Hishiyama et al., 1984].

From the PAN-based carbon fibers soaked in this dirt only for 5 min, a large number of colonies was cultured, more than 10 times the number with other fibers, indicating that some bacteria preferred to be with carbon fibers; in other words, carbon fibers attract some bacteria. In order to have such a large amount of clinging dirt, it was found that HT-type without sizing agents on their surface were preferable to HM-type and aeration of the solution was essential. During soaking of carbon fibers for a long time, carbon fibers were found to act as substrates for organism metabolism and are supposed to be able to construct a small ecosystem. Practical application of carbon fibers for water purification has successfully been performed in ponds, as shown in Fig. 4-41, and will be expanded to lakes, rivers and the sea [Kojima et al., 1997].

It was recently reported that some carbon fibers can sorb and retain hydrogen gas; more than 20 L(STP) of hydrogen gas per gram of carbon under a gas pressure of 12 MPa at 25°C [Chambers et al., 1998]. The fibers used were formed by catalytic decomposition of carbon-containing gases on small metal particles at temperatures over the range of 450–750°C and had the arrangement of carbon layers either perpendicular or at an angle with respect to the fiber axis, which the authors called "platelet" and "herringbone", respectively. The reported capacity of hydrogen adsorption of these fibers is over an order

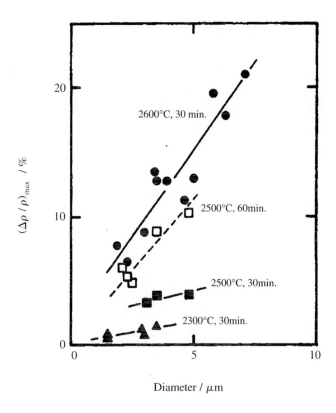

Figure 4-38 Dependence of $(\Delta\rho/\rho)_{max}$ on the diameter of the vapor-grown carbon fibers at different HTT and residence times [Ishioka et al., 1993c].

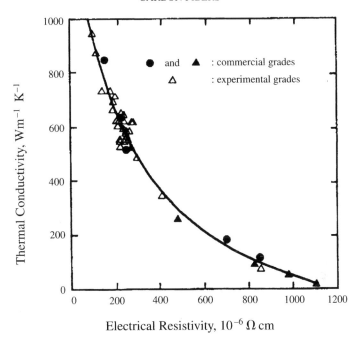

Figure 4-39 Relation between thermal conductivity and electrical resistivity on various mesophase-pitch-based carbon fibers [Lavin et al., 1993].

Table 4-7. Industrial applications of carbon fibers

Materials	Application	Related industries
Fibers		
	Insulation materials[a]	Electronics, automobiles, aircraft, atomic energy
	Sealing materials[a,b]	Chemicals, petrochemicals
	Absorption of bacteria[b]	Environmental
Composites		
With resin	Functional materials[a,b]	Electronics, communication, machines, automobiles, aircraft, chemicals
With carbon	Structural materials[b]	Sport, medical, space, aircraft, automobiles, communication, electrodes
With metal	Ablation materials[a,b]	Space, military
	Frictional materials[a,b]	Aircraft, automobiles, railways, machines
With ceramics	Cell electrodes[a,b]	Automobiles
With concrete	Construction materials[a,b]	Ships, buildings, housing, bridges, public works

[a]GP-grade, [b]HP-grade.

of magnitude higher than that found with conventional hydrogen storage systems, which is very much attractive for recent energy problems. Several investigators are interested in this result and are repeating the experiments. However, so far there has been no report of the reproduction of this high capacity.

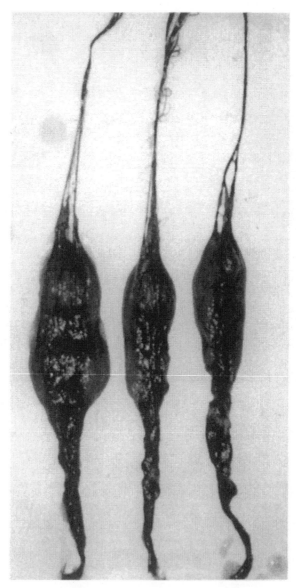

Figure 4-40 Clinging of microorganisms from the dirt onto a PAN-based carbon fiber strand after one day. [Courtesy of Prof. Kojima of Gunma National College of Technology.]

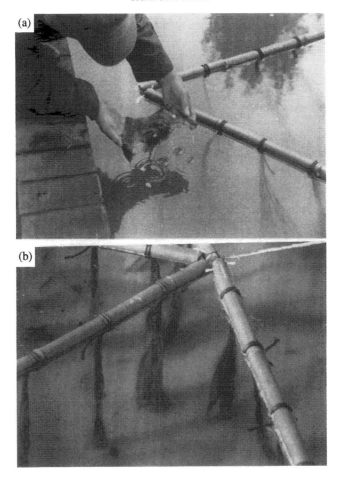

Figure 4-41 Use of carbon fibers in a pond with an area of 25 m² and a depth of 1 m: (a) after 30 min of hanging of the carbon fibers and (b) after 1.5 h with slight aeration. [Courtesy of Prof. Kojima of Gunma National College of Technology.]

Carbon multiwall nanotubes were successfully used for the cold sources of electron beams, to construct cathode-ray tube-type lighting elements and vacuum-fluorescence display panels, which promise to replace conventional thermoionic cathodes [Saito & Uemura. 1999].

CHAPTER 5

Porous Carbons

5.1. Formation of Pores in Carbon Materials

Almost all carbon materials contain pores, because they are the products of thermal decomposition of organic precursors, such as various resins and pitches. During their pyrolysis and carbonization, a large amount of decomposition gases is formed in a wide range of temperature, the profile of which depends strongly on the precursors. Figure 5-1 presents a general scheme of out-gas components and changes in residues from organic materials. The gas evolution behavior from organic materials is strongly dependent on the heating conditions, such as heating rate and pressure but, in common, hydrocarbon gases of aliphatic and aromatics come out at low temperatures and then foreign atoms, such as oxygen as CO_2 and CO, around 600°C, and at high temperatures mostly hydrogen, which has been understood to come from the condensed aromatic compounds grown in the preceding step (pyrolysis).

In the case of pitch, all constituent molecules are mobile at low temperatures and it has a sufficiently low viscosity to flow along movement of the bubbles of decomposition gases (thermoplastic nature). At high temperatures, however, the plasticity of the decomposed

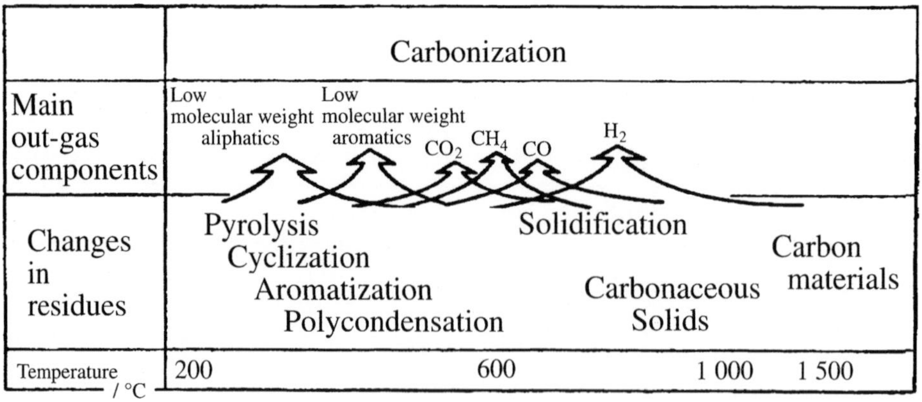

Figure 5-1 General scheme of out-gas components and changes in residues with carbonization.

pitches is lost because of condensation reactions and therefore out-gases leave pores behind. When the pitch has enough plasticity to flow at low temperatures or it is made to flow by other techniques, such as stirring, needle-like cokes are produced, which are important raw materials for classical carbon materials, graphite electrodes, and also some isotropic high-density graphites (Chapter 3). An example of such a flow pattern and elongated pores is shown in Fig. 5-2. These flow patterns are known to be due to the preferred orientation of polycondensed aromatic molecules and leads to carbon hexagonal layers after carbonization, and kept up to high temperatures. These elongated pores and the orientation of carbon hexagonal layers in the pore wall are known to absorb thermal shock very effectively.

Either by rapid heating, not allowing flow or by a slight oxidation like stabilization in the process of carbon fiber production (Chapter 6), porous carbon materials are prepared, of which pore sizes are usually widely distributed. Activated carbons with very high porosity and high specific surface area have been prepared by heat treatment of these porous carbons either with some additives, such as $ZnCl_2$ and KOH, or in an oxidative atmosphere, such as CO_2 and steam (activation) [Derbyshire et al., 1995].

In the case of phenol resin, there is no plasticity even at the beginning of thermal decomposition (thermosetting nature) and so large amounts of pores are formed after its carbonization. In most cases, pores are connected each other, i.e. open pores, because they are the paths by which decomposition gases go out of the residual carbons. If this thermal decomposition is done extremely slowly, enough to shrink the residues, most pores become unconnected, i.e. closed pores. The carbon materials produced through such a process with a very slow heating rate are called glass-like carbons (Chapter 3). Because

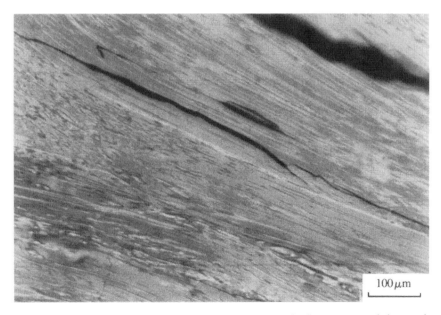

Figure 5-2 Optical micrograph of a needle-like coke, showing the flow pattern and elongated pore. [Courtesy of Dr Kakuta of Koa Petroleum Co.]

most pores are closed, it has low bulk density, but is gas impermeable. In this case, the precursors and preparation conditions are determining factors. A comprehensive model of the closed pores in glass-like carbons has already been shown in a previous chapter (Fig. 3-2). From small particles of phenol resin, such as spheres with a diameter of 10 μm, glass-like carbon spheres are obtained by a simple heat treatment in a flow of inert gas up to 1000°C, which have a very low specific surface area. Carbonization of the same phenol resin spheres in a flow of CO_2, however, gives a high surface area. In Fig. 5-3, the dependence of specific surface area on carbonization temperatures in different gas atmospheres is shown [Inagaki and Sunahara, 1998]. Up to 800°C, no effect of atmosphere during heat treatment on surface area is observed, but different areas are determined by using N_2 and CO_2, where the pyrolysis of phenol resin seems to be the main reaction. Above this temperature, the heat treatment atmosphere governs the surface area. Under high-purity argon gas, the surface area decreases with increasing HTT, which is supposed to be due to the shrinkage of pores by carbonization. In CO_2, however, the surface area increases abruptly owing to the formation of pores by oxidation (activation).

In carbon spheres heat-treated in N_2 up to 1000°C which show a very low specific surface area, a few $m^2 g^{-1}$, and in which it is difficult to measure pore size distribution by gas adsorption techniques, no pores are observed under FE-SEM. By scanning tunneling microscopy (STM), however, some ultramicropores with sizes between 0.5 and 0.9 nm can be detected, as shown in Fig. 5-4 [Vignal et al., 1999b; Inagaki et al., 1999].

The out-gas profile of a polyimide thin films, shown on a Kapton film with a thickness of 25 μm, is rather special, as shown in Fig. 2-13. At low temperatures, only CO and CO_2

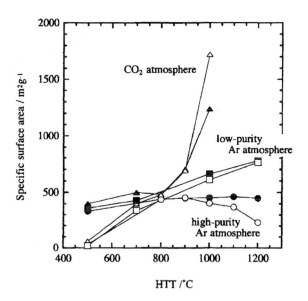

Figure 5-3 Dependence of specific surface area values measured using N_2 and CO_2 on heat treatment temperature (HTT) as a function of atmosphere during heat treatment [Inagaki and Sunahara, 1998]. Open marks show the surface area determined from N_2 adsorption and closed marks that from CO_2 adsorption.

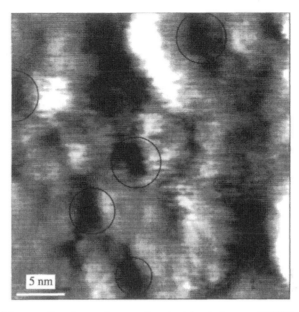

Figure 5-4 STM image of carbon sphere surface heat-treated up to 1000°C in N$_2$ atmosphere [Vignal et al., 1999b].

come out at a very narrow temperature range of 600–650°C and then mostly hydrogen gas and a small amount of methane are evolved. With this out-gassing, the film shrinks along the film surface by about 22% and along the thickness by about 70%. Even though this large and abrupt out-gassing is accompanied by pronounced shrinkage, very dense carbon films are obtained, with no pores or cracks. It was noted in Chapter 2 that these carbonized films could give highly crystallized graphite films. When thicker films of polyimide were employed as a precursor, the lower bulk density of carbon films and lower crystallinity of the resultant graphite films were obtained. Starting with the raw material polyamic acid, which was converted to polyimide through imidization, porous carbon films were prepared by applying the phase inversion method during the preparation of precursor polyimide films from polyamic acid solution [Hatori et al., 1992a, b], as will be shown later.

If plants such as trees and bamboos are selected as precursors for carbon, pores characteristic of plant texture are obtained. Some examples are shown in Fig. 5-5. This pore structure is a memory of the texture of the precursor plants. Recently, extensive work has been carried out on porous carbons from different woods reinforced by the impregnation of phenol resin, named "woodceramics" [Okabe et al., 1996a–c; Shibata et al., 1996].

Because structural units in carbon materials are parallel stacks of anisotropic hexagonal carbon layers, spaces with nanometric sizes and with slit-like shapes (nanospace or nanopores) are formed [Huttepain and Oberlin, 1990], which is one characteristic of some porous carbons. Some slit-shape nanospaces are reasonably supposed to be formed due to starting irregularity on structural units and also due to the cross-linkage between structural units [Kaneko, 1993]. In these slit-type nanopores, adsorbed molecules are expected to

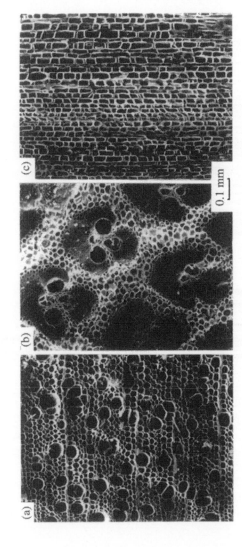

Figure 5-5 SEM micrographs of charcoal (a) and bamboo (b, c) carbonized at about 900°C in an inert atmosphere. [Courtesy of Prof. Hishiyama and Dr Yoshida of Musashi Institute of Technology.]

behave differently from those in large spaces and so various studies have been carried out [Kaneko, 2000]. From simulation based on the Grand Canonical Monte-Carlo method, CCl_4 molecules were expected to have a different structure in the nanospace with a slit width of 1 nm, which was experimentally proven by the measurements from small angle X-ray scattering [Suzuki et al., 1997]. A similar speciality of nanospace with slit-shape was reported on the adsorption on different adsorbates, formation of clathrates of hydrated NO in such a space [Kaneko et al., 1989; Fujiie et al., 1995], anomalous magnetic spin alignment of O_2 molecules at low temperatures [Kanoh and Kaneko, 1995], etc.

In carbon materials, particularly in graphite, a unique way of producing nanometric sized spaces is known, i.e. intercalation. As will be explained in Chapter 6, the intercalation compounds of graphite with alkali metals can provide nano-sized spaces for hydrogen and other hydrocarbon molecules. The schematic illustration of adsorbing hydrogen molecules into potassium–graphite intercalation compounds (K-GICs) is shown in Fig. 5-6. If some intercalation compounds are heated abruptly to high temperatures, such as 1000°C, the intercalates are decomposed and become vapors and, as a consequence, marked exfoliation of graphite occurs, as will be explained in Chapter 6, and its appearance will be shown in Fig. 6-26. This exfoliated graphite has a wide application after it is made into a thin sheet of graphite [Fujii and Dohi, 1986]. In industry, sulfuric acid intercalation compounds are usually employed and exfoliated to achieve about 300 times in volume.

5.2. Characterization of Pores

The pores in carbon materials are scattered over a wide range of size and shape, as discussed above. These pores are classified by their size, as illustrated in Fig. 5-7, usually into three classes: macropores (> 50 nm), mesopores (2–50 nm) and micropores (< 2 nm) according to IUPAC [Sing et al., 1985]. Micropores are further divided into super-micropores with a size of 0.7–2 nm and ultramicropores of less than 0.7 nm.

Pore size distribution in porous particles has been determined by different techniques, such as mercury porosimetry and gas adsorption isotherm measurement. Their fundamental theories, equipments, measurement practice and many results obtained so far have been reviewed in different publications [Rodriguez-Reinoso and Linares-Solano, 1995; Patrick, 1995]. The analysis of gas adsorption isotherms was first proposed by Langmuir based on single-layer adsorption and then modified by Brunauer et al. on

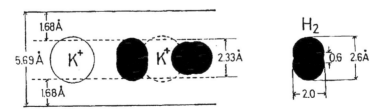

Figure 5-6 Nanospace in potassium–graphite intercalation compounds for hydrogen adsorption [Inagaki, 1988].

(a) Granular activated carbon

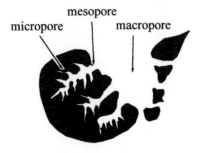

(b) Activated carbon fiber

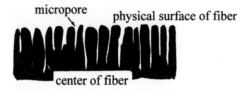

Figure 5-7 Pores in granular carbon materials (a) and in carbon fibers (b).

multilayer adsorption (so-called BET equation), Dubinin-Ashtokevic by assuming a distribution in adsorption potential, as a plot in order to give more detailed information on micropores, etc. A great effort has been exerted on the analysis of adsorption isotherms.

Recently, direct observation of pores on the surface of carbon materials was reported by using scanning tunneling microscopy (STM/AFM) [Donnet et al., 1994; Daulan et al., 1995] and a procedure for quantitative assessment of these pores was proposed [Vignal et al., 1999b; Inagaki et al., 1999]. In such surface observations, the problem is how to differentiate the pores from other surface defects, such as depressions and trenches. The following two criteria were proposed to measure the contour map around a pore:

(1) pore walls must be steep; their slope must be more than 1;
(2) pores must be so deep that the tip of the microscope cannot reach the bottom.

Figure 5-8 shows an STM image on the surface of carbon spheres which were prepared from phenol resin with heat treatment at 1000°C in N_2 atmosphere and had only a small number of ultramicropores, as shown in Fig. 5-4. Examples of a contour map around a pore and a trench are also shown. From observations on more than 150 pores, a pore size distribution was determined, as shown in Fig. 5-9. These STM observations gave only information on the entrance of pores, but is possible to show not only pores of a very small size but also the shape of their entrances.

On micropores, their fractal dimension was determined from the adsorption isotherm measurements using adsorbate molecules with different sizes [Avnir et al., 1983; Setoyama and Kaneko, 1995]. The number of molecules with size d, $N(d)$, which cover the pore surface, is expressed by the following fractal relation:

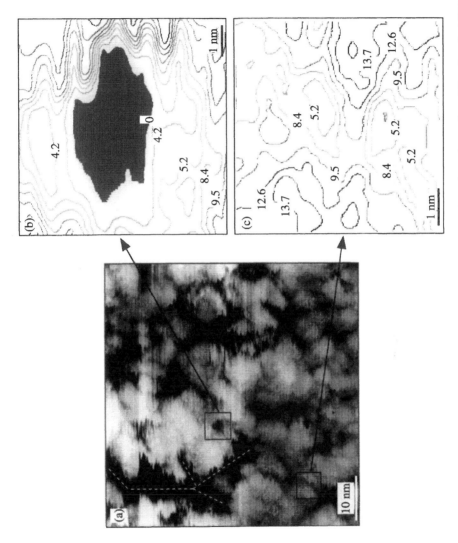

Figure 5-8   STM image (a); contour map around a pore (b) and a trench (c) on the surface of a carbon sphere [Vignal et al., 1999b].

$$N(d) \propto d^{-Da},$$

where Da is a fractal dimension for pore surface. Figure 5-10a illustrates the difference in the surface covering with molecules of different sizes. In Fig. 5-10b, the changes in fractal dimension with heat treatment are shown on two activated carbon fibers derived from different precursors. Da approaches 2.0, which corresponds to a smooth surface of pores, with increasing HTT.

By applying image analysis techniques to micrographs observed by STM, TEM, SEM or optical microscopy, depending on the size of the pores under study, the characteristics of pore structures are quantitatively expressed. Table 5-1 summarizes some results of the analysis on isotropic high-density graphite, of which optical micrographs were shown in Fig. 3-1 [Oshida et al., 1996].

By increasing the number of pores per mm², the porosity and averaged area per pore decrease, while the circularity of pores calculated from their area and periphery increases, suggesting that pores become round, but fractal dimension increases, indicating that the pore periphery becomes complicated. Changes in these parameters for pores are particularly pronounced for samples D–F in Table 5-1.

5.3. Control of Pores

Different pore sizes in carbon materials are required in their applications, examples of which will be shown later. Therefore, pore size distribution in carbon materials has to be

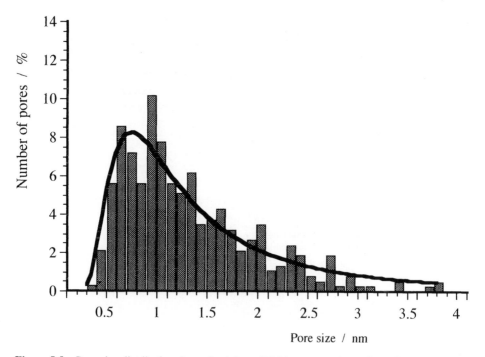

Figure 5-9 Pore size distribution determined from STM images on the surface of carbon spheres heat-treated in N_2 atmosphere with a simulated curve [Inagaki et al., 1999].

(a)   (b)

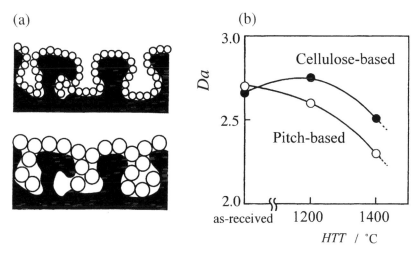

Figure 5-10 Determination of fractal dimension on the pore surface. (a) Scheme of adsorption of molecules with different sizes onto the pore surface; (b) changes in fractal dimension with HTT on two activated carbon fibers [Setoyama and Kaneko, 1995].

controlled during their preparation, selecting the precursor, carbonization and activation if necessary. Rather a broad distribution in pore size and shape is usually obtained in carbon materials, showing a sharp contrast to inorganic porous materials such as zeolite and MCM-41, which can achieve almost constant pore size. In order to compete in the adsorption performance with these porous inorganic materials and to use the advantages of carbon materials, such as high chemical stability, high temperature resistance and low weight, the control of pore size is essential. In order to control the pore size distribution in carbon materials, studies on the selection of precursors and preparation conditions have been extensively carried out and certain successes have been achieved [Sanada et al., 1992; Rodriguez-Reinoson and Linares-Salino, 1995; Patrick, 1995]. Porous carbon membranes were compared with other inorganic membranes for the application of gas

Table 5-1. Characteristics of pore structure in isotropic high-density graphites obtained from image analysis [Oshida et al., 1996]

Sample	Number of pores (mm^{-2})	Porosity (%)	Averaged area (μm^2)	Averaged circularity	Fractal dimension
A	414	21.9	545	0.65	1.40
B	446	25.3	506	0.69	1.40
C	480	21.1	395	0.66	1.33
D	822	21.7	244	0.68	1.46
E	1275	20.9	155	0.71	1.47
F	2731	12.0	31	0.79	1.56

Optical micrographs and mechanical properties of samples refer to Fig. 3-1 and Table 3-1, respectively.

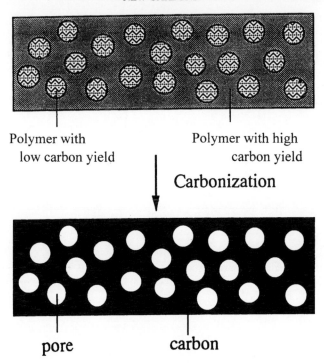

Polymer with   Polymer with high
low carbon yield   carbon yield

Carbonization

pore   **carbon**

Figure 5-11   Concept for the control of pore structure in carbon materials by the polymer blending method.

separation [Morooka and Kusakabe, 1999]. The control of pore structure in carbon materials was recently reviewed and discussed in detail [Kyotani, 1999].

The polymer-blending method, blending two kinds of polymers, one giving a high carbon yield and the other a very low yield, was recently developed to control the pores [Ozaki et al., 1997]. In Fig. 5-11, the concept of this method is explained. Figure 5-12 shows isotherms of N_2 adsorption/desorption for the carbons prepared at 600°C from a polyimide (PP) and from a blend of PP with polyethyleneglycol (PEG) [Hatori et al., 1992a, b, 1995a, b]. By blending PEG, which left pores in the carbon films from PP, a further increase in the adsorbed amount at high relative gas pressure (P/P_0) and pronounced hysteresis on the desorption are observed, being reasonably supposed to be due to the formation of mesopores in addition to micropores, which occur in the matrix carbon film derived from PP. By TEM observation, the presence of the mesopores with a diameter of about 10 nm was confirmed.

Porous carbon films were prepared from polyimide by using the technique of phase inversion during imidization of precursor polyamic acid [Hatori et al., 1995a, b]. The scheme of the process and SEM micrographs of a cross-section of the film as an example of pore structure are shown in Figs 5-13 and 5-14, respectively. Two advantages of this process for the preparation of porous carbon films are pointed out: the homogeneous size of pores can be controlled by selecting a mixing ratio between poor and good solvents for polyamic acid, and composite films of macroporous and dense microporous carbon layers can be prepared, as shown in Fig. 5-14b.

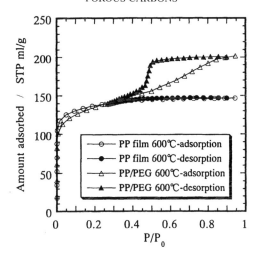

Figure 5-12 Isotherms of N_2 on the carbon films derived from a polyimide PP and from a blend of PP with polyethyleneglycol PEG, both of which are carbonized at 600°C [Hatori et al., 1992b].

5.4. Applications

(a) Activated carbon fibers

An important type of porous carbons which is widely used is activated carbons. The history of activated carbons goes back to the prehistoric era, when charcoal is known to be used for the purification of water and as a medicine. Granular activated carbons were prepared from different precursors and used in a wide range of industries. Their preparation, structure and applications are reviewed in different books [Sanada et al., 1992; Patrick, 1995; Marsh et al., 1997]. Fibrous activated carbons, activated carbon fibers, have been prepared recently and developed a new field of applications.

Activated carbon fibers have a number of advantages over granular activated carbons. The principal merit to prepare in fibrous morphology is its pore structure and a large physical surface area. Their pore structure is schematically shown in Fig. 5-7b, by comparing it with that of granular activated carbon (Fig. 5-7a), and the differences between activated carbon fibers and granular activated carbons are listed in Table 5-2. Granular activated carbons have different sizes of pores, macropores, mesopores and

Table 5-2. Comparison between activated carbon fibers and granular activated carbons

	Activated carbon fibers	Granular activated carbons
Size	10–20 μm in diameter	1–3 mm
Specific surface area ($m^2 g^{-1}$)	700–2500	900–1200
Physical surface are ($m^2 g^{-1}$)	0.2–2.0	~ 0.001
Average diameter of pores (nm)	< 40	–

micropores, whereas activated carbon fibers have mostly micropores on their surfaces. It is known that micropores are responsible for gas adsorption. In granular activated carbons, gas molecules always have to reach micropores by passing through macropores and mesopores, whereas in activated carbon fibers most micropores are exposed directly to gas. Therefore, the adsorption rate as well as the amount of adsorption of gases into activated carbon fibers are much higher than those into granular activated carbons. In Fig. 5-15, the adsorption and desorption behaviors of toluene are compared on these two carbons. Desorption of gas can be accelerated by raising the temperature on activated carbon fibers.

In Table 5-3, characteristics of commercially available activated carbon fibers prepared from different precursors are compared. A very high specific surface area up to 2500 m^2 g^{-1} and a high micropore volume up to 1.6 ml g^{-1} can be obtained in pitch-based carbon fibers. For the preparation of these carbon fibers with a very high surface area, such as 2500 m^2 g^{-1}, the precursors which give poorly crystalline carbons are recommended;

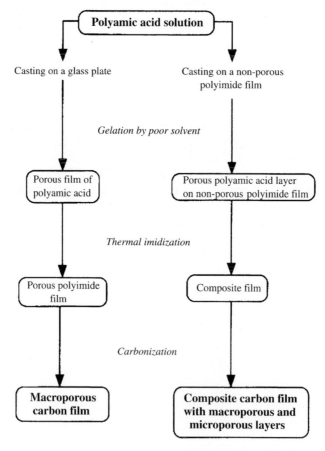

Figure 5-13 Preparation procedure of porous carbon films from polyimide by the phase inversion method [Hatori et al., 1992a, b].

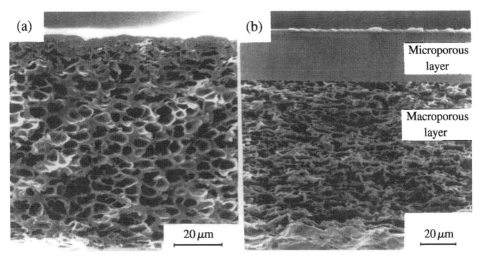

Figure 5-14 SEM micrographs of cross-section of porous carbon films prepared by the procedure of Fig. 5-13: (a) macroporous carbon film; (b) composite film of macroporous and microporous layers. [Courtesy of Dr Hatori of National Institute of Resources and Environmental Research.]

mesophase-pitch-based carbon fibers did not give a high surface area, whereas isotropic-pitch-based carbon fibers did [Wang and Inagaki, 1999]. New applications of these activated carbon fibers have been extensively developed, e.g. desulfurization in exhaust gases by oxidizing SO_2 to either sulfuric acid or sulfates on the surface of fibers [Kisamori et al., 1993; Mochida et al., 2000].

Another advantage of activated carbon fibers is the ability to prepare woven clothes and non-woven mats, which has been developed as a large application in small purification systems for city water and also as a deodorant in refrigerators, for example. In order to give fibers an antibacterial function and to increase their deodorant function, some trials on supporting minute particles of different metals, such as Ag, Cu and Mn, have been performed [Abe et al., 1996; Oya, 1997b].

Effective elimination of SOx from exhausted gases by using activated carbon fibers was recently reported [Mochida et al., 2000] and a new research project was started by the New Energy and Industrial Technology Development Organization (NEDO) in Japan.

Table 5-3. Some characteristics of commercially available activated carbon fibers

Precursor	Pitch	PAN	Phenol	Cellulose
Fiber diameter (μm)	10–18	7–15	9–11	15–19
Physical surface area ($m^2\,g^{-1}$)	0.2–0.6	0.9–2.0	1.0–1.2	0.2–0.7
Specific surface area ($m^2\,g^{-1}$)	700–2500	500–1500	900–2500	500–1500
Micropore volume (ml g^{-1})	0.3–1.6	–	0.22–1.2	–
Tensile strength (MPa)	100–200	200–370	300–400	60–100
Elastic modulus (GPa)	2–12	70–80	10–15	10–20
Tensile strain (%)	1.0–2.8	–2.0	2.7–2.8	–
Adsorption of benzene (%)	22–68	17–50	22–90	30–58
Iodine adsorption (mg g^{-1})	900–2200	–	950–2400	–

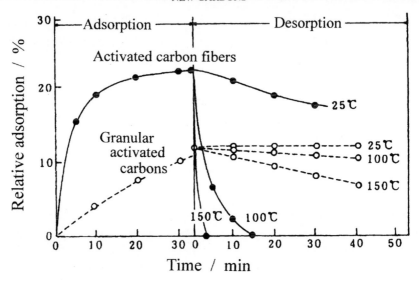

Figure 5-15 Adsorption and desorption of toluene vapor onto activated carbon fiber and granular activated carbon. [Courtesy of Dr Shimada.]

(b) Molecular sieving carbons

Molecular sieving carbons (MSCs) have a small pore size with a sharp distribution in a range of micropores, as shown in Fig. 5-16, compared with other activated carbons for gas

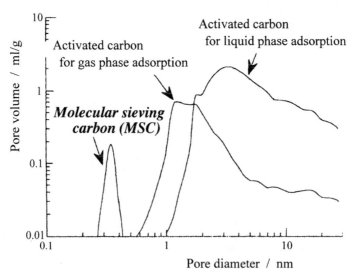

Figure 5-16 Pore size distribution of molecular sieving carbon (MSC), compared with activated carbons for gas and liquid phase adsorbates.

and liquid phase adsorbates. They have been used for adsorbing and eliminating unnecessary species with a very low concentration, ethylene gas adsorption to keep fruits and vegetables fresh, filtering of hazardous gases in power plants, etc.

However, the most important application of these MSCs is in gas separation systems (swing adsorption method). Since the pore size in MSCs is comparable to the size of adsorbate molecules, the adsorption rate of certain molecules, such as nitrogen, hydrogen and ethylene, depends strongly on the pore size of MSC; the adsorption rate of a gas becomes the slower for MSC with a smaller pore size. In addition, the temperature of adsorption governs the rate of adsorption of a gas because of the activated diffusion of adsorbate molecules in micropores; the higher the temperature the faster the adsorption. By controlling (swinging) these parameters, the temperature and pressure (i.e. concentration) of adsorbate gas, gas separation is performed. Swing adsorption methods are classified into two methods, temperature swing adsorption (TSA) and pressure swing adsorption (PSA) corresponding to which parameter is controlled.

Figure 5-17 shows the dependence of the relative adsorption ratio on time for oxygen and nitrogen. Adsorption of oxygen into an MSC reaches a maximum (equilibrium) within 5 min, but nitrogen is adsorbed less than 10% of equilibrium adsorption. From the column of MSC, therefore, nitrogen-rich gas comes out. On the desorption process oxygen-rich gas is obtained. By using more than two columns of MSC and repetition of these adsorption and desorption processes, nitrogen gas is isolated from oxygen. A flow diagram of PSA for the separation of nitrogen from air is shown in Fig. 5-18. This swing adsorption method, practically PSA, for gas separation has the advantages of low energy cost, room temperature operation, compact equipment, etc.

(c) Porous carbons for car canisters

There is a strong demand to reduce and control the composition of exhaust gas from cars because of their increase in number. Cars are a major source of global warming gases. In order to treat these exhaust gases, different catalysts have been used and are being studied extensively. Recently, an activated carbon has been used for adsorbing gasoline vapor in the tank of a car during parking and using it during running.

The flow of gasoline vapor in a car is illustrated in Fig. 5-19. Even during parking, gasoline vaporizes, particularly in hot weather. Some statistics show that most cars are

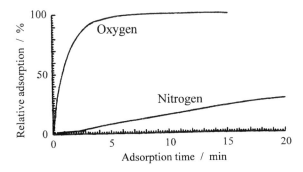

Figure 5-17 Changes in relative adsorption of oxygen and nitrogen onto molecular sieving carbons with time.

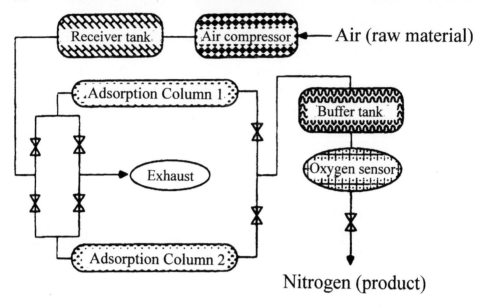

Figure 5-18 Flow diagram for nitrogen separation system by pressure swing adsorption (PSA) process.

parked for the whole day, with the running time usually being very short in comparison with the parking time. The gasoline vapor formed during parking should be collected in a canister consisting of activated carbons, in order to save gasoline and to avoid contamination of the air. The adsorbed gasoline is desorbed during running by introducing air and transferred to the engine.

For this purpose, porous carbons have to have rather large pores, as shown in Fig. 5-20, compared with activated carbons for general purposes, because gasoline vapors are mostly adsorbed as liquid due to their capillary condensation and their surface tension in the pores of carbons.

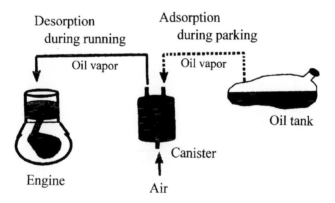

Figure 5-19 Flow of gasoline vapor in a car.

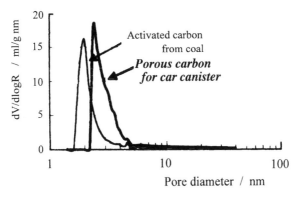

Figure 5-20   Pore size distribution in porous carbon for a car canister, compared with other activated carbons.

(d) Porous carbons for electric double-layer capacitors

At the interface between two different phases, in the present case between a solid electrode and a liquid electrolyte, a very thin layer is formed in which two charges, positive and negative, are aligned as illustrated in Fig. 5-21, an electric double layer. By using porous carbon materials with a very high surface area in both positive and negative electrodes, a large amount of electrical charge was found to be stored and electric double-layer capacitors were developed [Yoshida et al., 1987, 1988; Nishino, 1988]. The fundamental concept of this capacitor is illustrated in Fig. 5-21. The total amount of electric charges aligned in double layers on both electrodes increases by the application of potential difference and it is easily understood to depend on the area of this interface, i.e. the surface area of solid electrodes. The large specific surface area of activated carbons, which cannot be obtained in other materials, as well as their electrical conductance, are used in electric double-layer capacitors. Particularly by using activated carbon fiber cloths prepared from

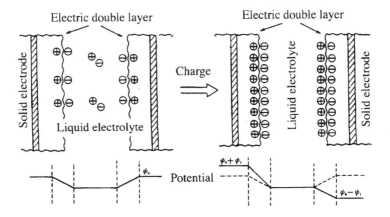

Figure 5-21   Fundamental scheme of an electric double-layer capacitor [Nishino, 1988].

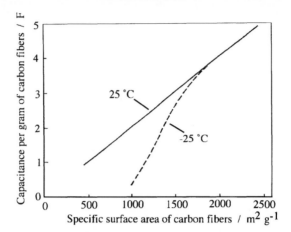

Figure 5-22 Dependence of capacitance per gram of activated carbon fibers on their specific surface area at room temperature and − 25°C [Nishino, 1988].

phenol resin and organic electrolytes such as propylenecarbonate with tetraethyl-ammonium perchlorate, small capacitors with a coin-shape were realized, with a high capacitance, high working voltage and high reliability [Tanahashi et al., 1990]. In Fig. 5-22, capacitance per gram of activated carbon fibers in a capacitor is plotted against specific surface area of the activated carbon fibers used. At room temperature, a linear relation is observed, revealing that a high surface area is desired for high capacitance. At low temperatures, carbon fibers with a relatively low surface area do not give capacitance.

The discharge behavior of a capacitor using activated carbon fibers is shown in Fig. 5-23; an extremely long time discharge is possible if the discharge current is small enough,

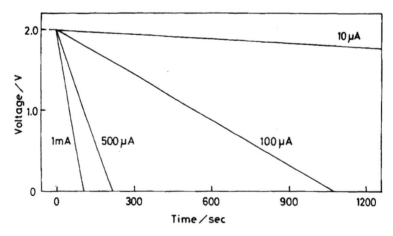

Figure 5-23 Discharge curves of a coin-type electric double-layer capacitor using activated carbon fibers at room temperature as a function of discharge current [Tanahashi et al., 1990].

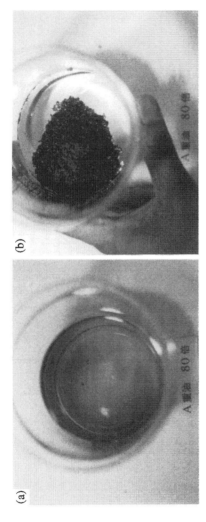

Figure 5-24 Sorption of A-grade heavy oil onto exfoliated graphite. (a) Floating heavy oil on water before adding exfoliated graphite, and (b) 1 min after adding exfoliated graphite [Toyoda et al., 1999].

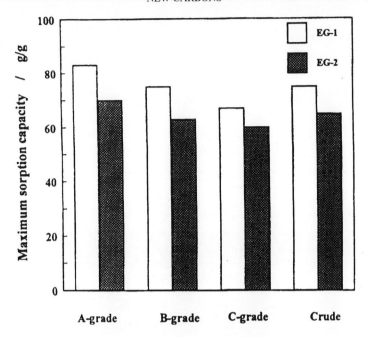

Figure 5-25   Maximum sorption capacity of two exfoliated graphite (EG-1 and EG-2) with different bulk density for different grades of heavy oils [Toyoda et al., 1999].

smaller than 10 μA, but the discharge voltage decreases quickly with time if the current is larger than 100 μA. This discharge behavior of the capacitors was not affected by ambient temperature in relatively wide range, from –25 to +85°C. There are certain advantages over secondary batteries, such as Ni–Cd batteries, including the possibility of semipermanent repetition of charge–discharge cycles, very short charging time and very simple circuit for charging, even though the discharge current is limited to a small value.

Based on these characteristics, electric double-layer capacitors are used for back-up source of memory in computers and also used together with small batteries. In particular, its use together with solar batteries expands markedly the application possibility of solar batteries due to the increase in the stability of the discharge voltage and the life. For the back-up of large memories, capacitors of either round or square cylindrical type with a diameter of 7–13 mm and length of 25 mm, for example, have been commercialized, in which granular activated carbons were mostly used. For small electronic devices, coin-type capacitors using activated carbon fibers with a size of 9.5 mmΦ and 2.1 mm high, for example, were used.

(e) Exfoliated graphite

Exfoliated graphite has been prepared on an industrial scale by abrupt heating of intercalation compounds of natural graphite flakes with sulfuric acid in most cases to a temperature as high as 1000°C and used for sealing as graphite sheets after compressing and rolling [Fujii and Dohi, 1986]. Therefore, they were never used as porous materials.

Recently, however, it was found that this exfoliated graphite can sorb large amounts of heavy oil [Cao et al., 1996; Toyoda et al., 1998a, b]. The photographs in Fig. 5-24 show that a lump of exfoliated graphite can sorb a large amount of A-grade heavy oil very quickly, within 1 min. The characteristic brown color of a heavy oil floating on water disappears quickly after adding a lump of exfoliated graphite. Heavy oil seems to be attracted to exfoliated graphite and the shape of the lump is maintained even after sorption of a large amounts of heavy oil. The maximum sorption capacity was determined from these observations. The results are shown in Fig. 5-25 on two exfoliated graphite having slightly different bulk density and on four different grades of heavy oil [Toyoda et al., 1999]. An exfoliated graphite (EG-1), which has a bulk density of about 0.006 g cm^{-3}, has a maximum sorption capacity of 86 g of A-grade heavy oil per 1 g of exfoliated graphite, but slightly lower capacity for C-grade heavy oil, which has a much higher viscosity at room temperature. Another exfoliated graphite (EG-2), which has a slightly higher density of 0.01 g cm^{-3}, shows a lower capacity of all four heavy oils. The compression of this exfoliated graphite, i.e. the increase in bulk density, was found to decrease the sorption capacity of all heavy oils [Toyoda et al., 1999]. Recent experimental results on sorption and recovery were reviewed by Toyoda and Inagaki [2000].

CHAPTER 6

Intercalation Compounds

6.1. Structural Characteristics

It is well known that graphite can accept various atoms, ions and even molecules between its interlayer space (gallery) of hexagonal layers of carbon atoms. The phenomenon of such an insertion is called intercalation and the products of this intercalation reaction are usually called graphite intercalation compounds. These intercalation compounds are classified into a member of the host–guest compounds, where the host is graphite with a layer structure and the guest is an intercalate. The properties of host graphite are strongly modified by the intercalation; for example, the intercalation compound of potassium up to a composition of KC_8 (stage-1 structure) has a golden color and that to KC_{24} (stage-2) blue, while the compound with fluorine $(CF)_n$ is white. However, even carbon materials, in which the graphite structure is not developed and crystallite sizes along both a- and c-axes are small, can be intercalated by some intercalates, such as potassium. In order to cover all of these compounds, from those of highly crystallized graphite to those of low crystallinity carbons, therefore, the word "graphite intercalation compounds" (GICs) is not appropriate and so the term "intercalation compounds" (ICs) was used. The effect of host carbon materials on intercalation reactions has just been started to be investigated. These ICs have many possibilities for novel applications [Inagaki, 1987, 1989], including the recent remarkable development of lithium ion rechargeable batteries. Hereafter, "ICs" is mostly used to express a whole group of compounds, including GICs, but if the host is clearly known to be graphite "GICs" is used.

In Fig. 6-1, the structure of $NiCl_2$-IC is schematically shown in order to explain the principal characteristics of the compounds. A characteristic of ICs is the charge transfer between intercalates and carbon layers, which is the main cause of different functions of ICs, as will be seen below. This charge transfer can occur both ways, from carbon layers to intercalate and vice versa, because carbon, particularly graphite, has two kinds of carriers, negative electrons and positive holes, in its band structure. In other layered compounds such as clays and transition metal sulfides, this bimodal intercalation phenomenon has never been observed; for example, clays can accept only positive ions such as sodium. This is the main reason why the intercalates for carbon are so numerous. Some representative intercalates are tabulated in Table 6-1, where the intercalates are classified on the basis of the direction of charge transfer between intercalates and carbon

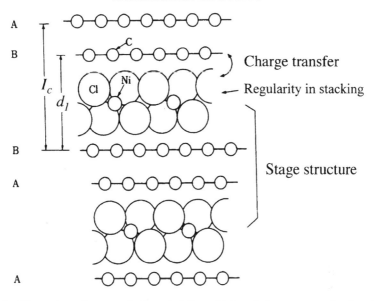

Figure 6-1 Illustration of structural characteristics of intercalation compounds of carbon (ICs), showing stage-2 $NiCl_2$-IC.

layers; donor-type intercalates which give electrical charge to carbon layers and become positive ions in the gallery of carbon layers, and acceptor-type intercalates which receive electrical charge from carbon layers and become negative ions in the gallery. This charge transfer between intercalate and carbon does not occur at 100%; in other words, the ionization ratio of intercalates in the carbon gallery is not always 100%. In general, donor-type intercalates such as alkali metals have a high ionization ratio in ICs, close to 100%,

Table 6-1. Representative intercalates for intercalation compounds

Bond nature	Electronic state of intercalate	Examples of intercalates
Ionic	Donor type	Li, Na, K, Rb, Cs
		Ca, Sr, Ba
		Mn, Fe, Ni, Co, Zn, Mo
		Sm, Eu, Yb
		K-Hg, Rb-Hg
		$K-NH_3$, $Ca-NH_3$, $Eu-NH_3$, $Be-NH_3$
		K-H, K-D
		K-THF, $K-C_6H_6$, K-DMSO
Ionic	Acceptor type	F, Br, ICl, IBr, IF_5
		$FeCl_2$, $FeCl_3$, $NiCl_2$, $AlCl_3$, $SbCl_5$
		AsF_5, SbF_5, NbF_5, XeF_5
		CrO_3, MoO_3
		HNO_3, H_2SO_4, $HClO_4$, H_3PO_4
Covalent		F, O(OH)

but acceptor-type ones have rather low ionization ratios of 10–30%. In acceptor-type ICs, therefore, there are neutral species, in Br-ICs Br_2 molecules intercalating together with Br^- and in sulfuric acid-ICs H_2SO_4 with HSO_4^-. In these donor and acceptor types of ICs, the carbon layers are kept as flat as in the host carbons.

Another characteristic of ICs is the stage structure, which can be in a wide range from 1 to more than 10. The stage number n defines how many carbon layers are located between two intercalate layers, as shown in Fig. 6-2. In a stage-1 structure, every gallery of carbon is filled with intercalates, the carbon and intercalate layers being stacked alternatively along the normal to these layers, and therefore one carbon layer between every two intercalate layers. There are two carbon layers between the intercalate layers in stage-2, three carbon layers in stage-3, and so on. This stacking regularity between carbon and intercalate layers (staging) has been observed for various intercalates and their stage number n has been reported to reach 10 and more. Such a wide range of staging as in ICs of carbons has never been observed in other intercalation compounds of clays, metal sulfides, etc. This wide range of staging leads to a wide range of composition of ICs and, as a consequence, a wide range of functions.

In the case of graphite, it has to be pointed out that the stacking sequence between two intercalated graphite layers changes to AA in most cases, although pristine graphite has an ABAB stacking sequence, and the graphite layers that are not intercalated keep an ABA stacking. The interlayer spacing between these non-intercalated graphite layers is believed to keep that of host graphite, i.e. 0.3354 nm, although the spacing between intercalated graphite layers depends primarily on the size of intercalate. In Fig. 6-3, the spacing between intercalated graphite layers, d_I in Fig. 6-2, by different metal ions is plotted against the crystal radius of the metal, which exhibits a strong dependence of interlayer

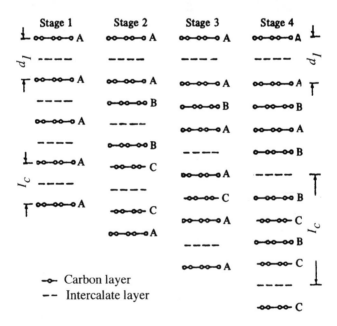

Figure 6-2 Stage structure of intercalation compounds.

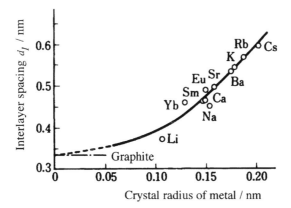

Figure 6-3 Relation between the spacing d_I in ICs of various metals and crystal radius of metals [Inagaki, 1978].

spacing d_I on the size of intercalates. In the case of carbon materials which have turbostratic stacking and larger spacing than 0.3354 nm for graphite, however, there have been no studies on the stacking regularity between two carbon layers and a sandwiched intercalate layer, or on the spacing for a non-intercalated gallery, whether it retains the original value larger than 0.3354 nm or not.

This stage structure of ICs is determined from their XRD patterns. In Fig. 6-4, XRD patterns of $MoCl_5$-GICs with stage-1, -2 and -4 structures, where only a series of $00l$ diffraction lines is detected, because of the strong orientation of layer planes parallel to the holder surface for an X-ray diffractometer. The index l is determined from the observed values of spacing, d_{obs}, on each diffraction line and the following equation:

$$I_c = d_I + 0.3354 \times (n-1) = l \times d_{obs},$$

to give a unique value of identity period I_c (refer to Fig. 6-2) for each compound.

By using a common d_I value of 0.922 nm, stage-1, -2 and -4 structures for three $MoCl_5$-GICs in Fig. 6-4 are decided, which have I_c-values of 0.922, 1.257 and 1.927 nm, respectively. The I_c-value thus determined is only the distance of repeating periodicity of the layers of carbon and intercalate along their normal, not necessarily the lattice constant of the crystal, this being the reason why I_c has been called the identity period. In stage-1 K-GIC, the layer of potassium intercalates is known to stack with the mutual relation of $\alpha\beta\gamma\delta$ and so the crystallographic lattice constant must be four times the I_c-value, which is determined from the X-ray powder pattern.

In ICs with most transition metal chlorides, the intercalate layers are composed of octahedra of chloride ions with metal ions at the center which are shearing edges, and so the interlayer spacing for intercalated gallery d_I does not depend on metal species. In these GICs, three layers of atoms, chloride–metal–chloride, are formed, as can be seen from Fig. 6-1 on $NiCl_2$-IC.

ICs with two different intercalates were synthesized, which were called ternary ICs, i.e. two different intercalates and carbon. In these ternary ICs, various intercalate layers in the carbon gallery are formed, bi-intercalation, completely mixed, islands-formed, triple-layer and coordinated types, as shown schematically in Fig. 6-5.

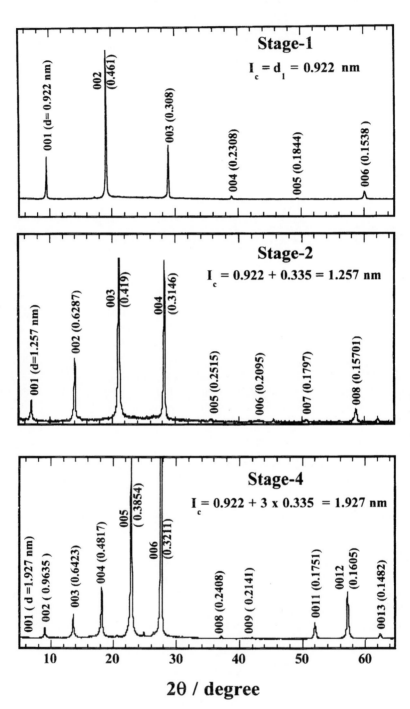

Figure 6-4 Illustration of the determination of stage number n and identity period I_c in ICs from the X-ray diffraction pattern, using MoCl$_5$-GICs.

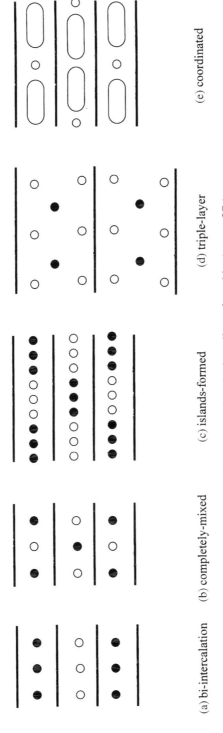

(a) bi-intercalation   (b) completely-mixed   (c) islands-formed   (d) triple-layer   (e) coordinated

Figure 6-5   Models for the arrangement of two different intercalates in the gallery of graphite (ternary ICs).

Ternary ICs with different couples of intercalates were reported to have a bi-intercalation type structure. When stage-2 $FeCl_3$-GIC reacts with $AlCl_3$, for example, $AlCl_3$ molecules intercalate preferentially into the gallery where no $FeCl_3$ is present, and so ternary GIC is formed, with a bi-intercalation type structure, graphite–$FeCl_3$–graphite–$AlCl_3$–graphite repeating periodicity [Inagaki and Ohira, 1993]. When stage-1 $FeCl_3$-GIC was reacted with $AlCl_3$, however, $AlCl_3$ molecules replaced partly $FeCl_3$ in the gallery of graphite and resulted in completely mixed-type ternary compounds. Figure 6-6 shows electron density distribution curves along the c-axis, which are determined from X-ray powder patterns with $00l$ diffraction lines by Fourier synthesis on these two types of GICs.

An example of a ternary GIC with a triple-layer type structure is $KC_8H_{2/3}$, as shown schematically in Fig. 6-7a, which is synthesized by the reaction between stage-1 K-GIC with hydrogen gas and has a stage-2 structure [Furdin et al., 1976]. In this compound, potassium forms double layers in the gallery of graphite and hydrogen is located in between these two potassium layers. On this $KC_8H_{2/3}$ compound additional alkali metals were reported to be able to intercalate and form a bi-intercalation type structure (Fig. 6-7b) [Lagrange et al., 1974]. When graphite powder was kept in tetrahydrofuran (THF) solution of alkali metals at room temperature, ternary compounds of alkali metal and THF molecule with a coordinated-type structure were formed [Nomine and Bonnetain, 1969; Tanaike and Inagaki, 1997], where either the complexes of Li with tetrahedrally coordinated THF or the K-THF complexes deformed to a rather flat shape were intercalated, as shown in Fig. 6-8.

A variety of relative positions of intercalates to the hexagons in carbon layer was determined in mostly binary ICs. Alkali metals are located at registered positions to carbon hexagon in the different ways shown in Fig. 6-9, which correspond to the compositions of MC_8, MC_6, MC_4, MC_3 and MC_2 if a stage-1 structure is assumed. MC_8 is realized in ICs with heavy alkali metals K, Rb and Cs in the stage-1 structure and MC_6 with Li. In these MC_8 and MC_6, all metal ions are located just above the center of the hexagons of the carbon layer (registered) and the distance to nearest neighbor ions is a multiple of the lattice constant of graphite a_o (commensuration), $2a_o$ (2×2 superstructure) and $\sqrt{3}a_o$ ($\sqrt{3} \times \sqrt{3}$ superstructure), respectively.

In ICs of heavy alkali metals with a stage structure greater than 2, the potassium arrangement in the intercalate layer does not have a simple relation with carbon layers. At a temperature between room temperature and 125 K, potassium ion layer in the gallery have a random arrangement and, between 125 and 95 K, their arrangement is discommensurate with the hexagonal lattice of the carbon layers and below 95 K commensurate structure with a regularity of $\sqrt{7} \times \sqrt{7}$ R19.1°, as shown on stage-2 K-GIC of the composition KC_{24} in Fig. 6-10.

The ICs with alkali metals with higher compositions than MC_6 and MC_8 were synthesized under high pressures [Nalimova et al., 1992]. A detailed study on LiC_2 formed under high pressure showed that some Li–Li bonds in the intercalate layers are smaller than those in metal, but that the charge transfer to carbon layers is comparable to that in LiC_6, suggesting a certain degree of covalent bonding.

On stage-4 $MoCl_5$-GIC, bright dots overlapping on the hexagonal pattern of carbon atoms in STM images were observed, as shown in Fig. 6-11a, indicating the presence of

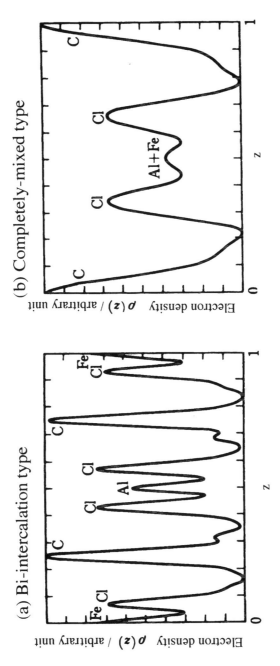

Figure 6-6 Electron density distribution along the c-axis in stage-1 $FeCl_3$–$AlCl_3$–ICs with bi-intercalation and completely mixed types [Inagaki and Ohira, 1993].

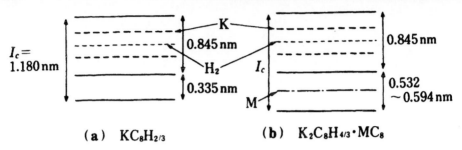

(a)   KC$_8$H$_{2/3}$ **(b)   K$_2$C$_8$H$_{4/3}$·MC$_8$**

Figure 6-7   Structure models for ternary ICs with triple-layer type [Furdin et al., 1976; Lagrange et al., 1974].

a superstructure due to chloride ions under the carbon layer, and the structure of the layer of the intercalate MoCl$_5$ was proposed by assuming di-molecules as Fig. 6-11b, where chloride ions are not registered and a discommensurate arrangement of MoCl$_5$ is shown [Vignal et al., 1999a]. In FeCl$_3$-GICs, no relation in the arrangement of FeCl$_6$ octahedra along the layer to the graphite layer was reported.

   In addition to two types of intercalates, donor- and acceptor-type, fluorine and oxygen can make covalent bonds with carbon atoms in hexagonal carbon layers (Table 6-1). The structure and chemical composition of the compounds with fluorine are well understood, in which the graphite layers are no longer flat but the structures corresponding to stage-1 and -2 were synthesized with chemical compositions of CF and C$_2$F [often written as (CF)$_n$ and (C$_2$F)$_n$ in the references], of which structural models are shown in Fig. 6-12 [Touhara et al., 1984a]. In (C$_2$F)$_n$, half of the carbon atoms form covalent bond with fluorine and the other half are also covalently bonded with carbon atoms in neighboring layers. It was

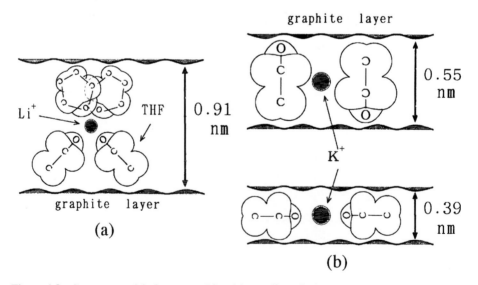

Figure 6-8   Structure models for ternary ICs with coordinated type.

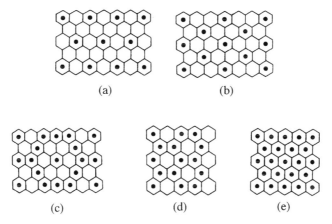

Figure 6-9 Position of alkali metals related to hexagonal carbon layers: (a) MC_8, (b) MC_6, (c) MC_4, (d) MC_3, (e) MC_2.

found that when fluorine reacts with graphite at a relatively low temperature using a catalyst, fluorine can intercalate into the graphite gallery with ionic interaction, i.e. without covalent bonding, by keeping the graphite layers flat, as will be explained later.

It was pointed out by many authors that a simple model for stage structure, as in Fig. 6-2, cannot be explained by the stage transition phenomena during either formation or decomposition of various ICs. For instance, the transition from stage-3 to -2 during formation requires the movement of intercalates already presented in the gallery to the neighboring gallery. A model, shown in Fig. 6-13, has been proposed [Daumas and Herold, 1969], where the intercalates are located in every gallery of graphite but make domains with a stage structure. In this model, the stage transition can be explained only by the diffusion of intercalates in the same gallery and so has been accepted by many authors. However, clear direct evidence for this model has not yet been presented. It presents difficulty in interpreting the formation reactions of bi-intercalation type structures; in the reaction of stage-2 $FeCl_3$-GICs with $AlCl_3$ described above (Fig. 6-6), for example, $AlCl_3$ has to pass by the intercalate layer of $FeCl_3$ in order to reach the non-intercalated region to form a bi-intercalation structure.

The STM observation of $MoCl_5$-GIC shows clear-cut boundaries between small intercalated domains and non-intercalated ones, as shown in Fig. 6-14b, but some

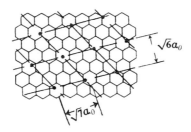

Figure 6-10 Position of potassium atoms relative to the hexagonal carbon layer in stage-2 KC_{24}.

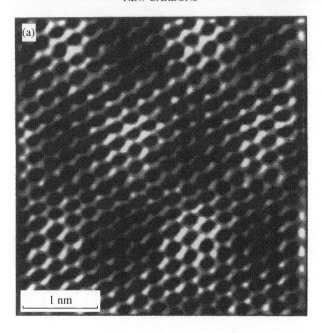

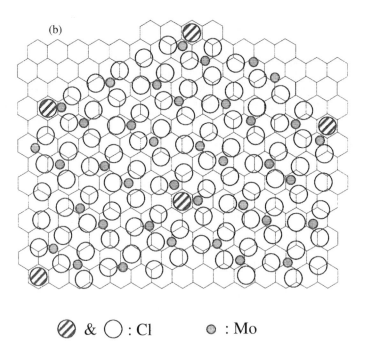

⊘ & ◯ : Cl   ⊙ : Mo

Figure 6-11   STM image showing the presence of a superstructure in stage-4 $MoCl_5$-GIC (a) and a model of its structure (b) [Vignal et al., 1999a].

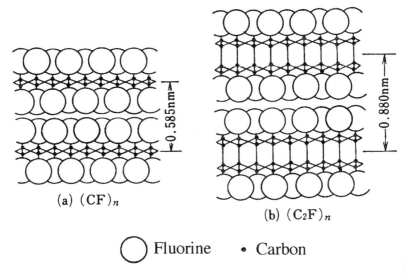

(a) $(CF)_n$

(b) $(C_2F)_n$

◯ Fluorine   • Carbon

Figure 6-12   Structure models for $(CF)_n$ and $(C_2F)_n$ [Touhara et al., 1984a].

corrugation at the boundaries for large intercalated domains, as in Fig. 6-14a [Vignal et al., 1999a]. In the former, the alignment of carbon atoms seen on the STM image was not disturbed at the boundary.

6.2. Synthesis

Various methods have been employed to synthesize ICs. The methods for the synthesis of ICs are summarized in Table 6-2, where they are classified based firstly on whether the host carbon is separated from or directly in contact with the intercalate source, and secondly on the state of intercalates during reactions.

—— Carbon layer
······ Intercalate layer

Figure 6-13   Daumas–Herold model for stage structure in ICs [Duamas and Herold, 1969].

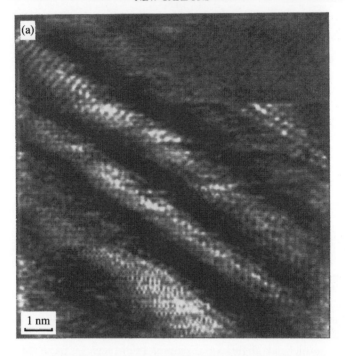

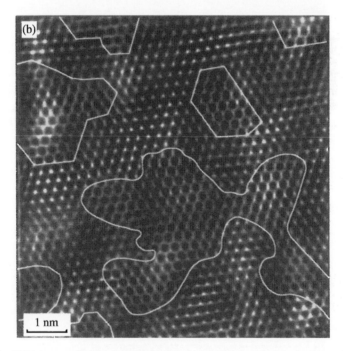

Figure 6-14 STM images for the boundaries between intercalated and non-intercalated regions. (a) Boundary with a large intercalated domains; (b) boundary with small intercalated domains (trimmed by white line) [Vignal et al., 1999a].

Table 6-2. Synthesis methods for intercalation compounds

Mutual position between host and guest	State of intercalate	Motive force for reaction	Examples of intercalates
Separated (two-bulb method)	Vapor	Thermal	K, $FeCl_3$
Contacted (mixing method)	Solid powder	Thermal	$FeCl_3$, $MoCl_5$
	Liquid	Thermal	Br_2
	Alloy	Thermal	Li–Na
	Molten salt	Thermal	$CuCl_2$–KCl
	Aqueous solution	Chemical	H_2SO_4, HNO_3
		Electrochemical	H_2SO_4,
	Non-aqueous solution	Electrochemical	Li, Li-DMSO
		Chemical	K-THF, $FeCl_3$-DME
		Light-assisted	$FeCl_3$-$CHCl_3$

DMSO: dimethylsulfoxide, THF: tetrahydrofuran, DME: dimethylethane.

The two-bulb method has been used to synthesize various ICs, where the host carbon and intercalates are separated in respective bulbs, as shown in Fig. 6-15a. The intercalation reaction, therefore, proceeds due to the reaction between solid carbon and intercalate vapor, the pressure of which is controlled by the temperature of the intercalate source T_i. The temperature of host carbon T_h is usually kept a little higher than T_i, in order to avoid the deposition of intercalate vapor on the surface of ICs formed. The stage structure can be controlled by the vapor pressure of the intercalate, practically the temperature difference $T_h - T_i$. Figure 6-15b shows an isotherm for potassium intercalation by plotting the composition of the compound against the temperature difference $T_h - T_i$. This method has some advantages, i.e. the ability to obtain a homogeneous stage structure and to introduce fewer defects in the resultant ICs because of the slow reaction rate. However, some disadvantages are also pointed out, e.g. the difficulty in synthesizing a large amount of compounds, and the long time needed for the reaction.

The methods for the synthesis of ICs where host and intercalate are in direct contact (mixing method) are classified on the basis of the state of intercalates, solid and liquid (Table 6-2). The first synthesis of Li-GIC with a stage-1 structure, LiC_6, was reported to be achieved from a mixture of metallic lithium with host graphite under pressure [Guerard and Herold, 1975]. Mixtures of transition metal chlorides with the host carbon at a temperature lower than their melting point can also give ICs. An example of the intercalation reactions with solid $CuCl_2$ will be compared with other methods later (see Fig. 6-17). In this method, however, the practical species reacting with host carbon must be the vapor of intercalates, although its pressure is low.

The liquid state of intercalates can be either their own melts, their molten salts with other species which do not intercalate into graphite, or their solutions. Bromine is in liquid phase at room temperature and so Br-ICs can be synthesized easily by direct dipping the host carbon into bromine. The molten state of intercalates which has been used for the synthesis of ICs is either a molten alloy of Li–Na or molten salts of transition metal chlorides with KCl. Using Li–Na molten alloy, Li-GICs were obtained at a temperature

(a)

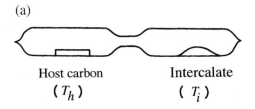

Host carbon   Intercalate
(T_h)   (T_i)

(b)

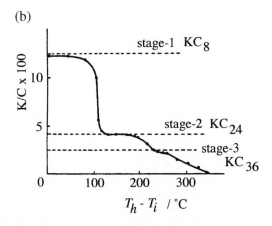

Figure 6-15 Two-bulb method for the synthesis of ICs (a) and an isotherm for potassium intercalation (b).

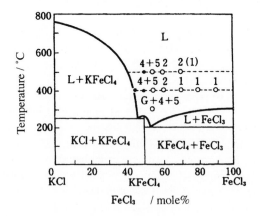

Figure 6-16 Stage number of GICs formed in molten salts of FeCl$_3$–KCl system with different compositions [Wang and Inagaki, 1992].

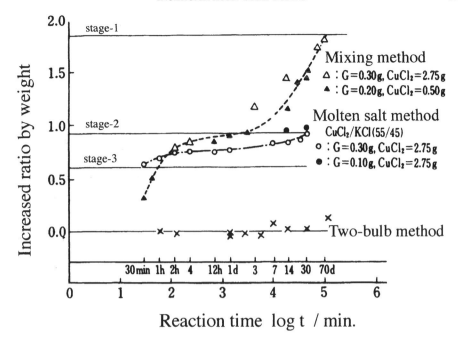

Figure 6-17 Comparison among two-bulb, solid mixing and molten salt methods for the synthesis of CuCl₂–GICs at 380°C [Wang and Inagaki, 1990].

lower than the formation temperature of lithium carbide [Basu et al., 1979]. Various transition metal chlorides were quickly intercalated at relatively low temperatures by using their molten salts, e.g. the FeCl₃–KCl system. In Fig. 6-16, the stage number of GICs formed is shown in the phase diagram of the FeCl₃–KCl system, where KCl lowers the melting point of the system and changes the chemical activity of intercalate FeCl₃, although it cannot intercalate into the graphite gallery [Inagaki and Wang, 1992; Wang and Inagaki, 1992].

In Fig. 6-17, two-bulb, molten salt and solid mixing methods are compared for the CuCl₂ intercalation reaction at 380°C [Wang and Inagaki, 1990]. Since the reaction temperature of 380°C is lower than its melting point and its vapor pressure is so low, the two-bulb method cannot produce GICs. In the molten salt of CuCl₂–KCl, intercalation proceeds so rapidly at the beginning that a stage-3 structure is formed within 1 h. However, only the stage-2 structure is obtained even after a long reaction time, probably because of insufficient chemical activity of CuCl₂ in the molten system to reach a stage-1 structure. The direct mixing method leads to the formation of a stage-1 structure, although it takes rather a long time because of the low vapor pressure below its melting point.

In order to intercalate MoCl₅ into graphite, it was known to be sufficient that a mixture of reagent-grade MoCl₅ powder with host graphite, either powder or film, was sealed in an ampoule under vacuum and heated to a temperature between 150 and 300°C. In contrast, AlCl₃ can be intercalated into graphite only under the co-existence of chlorine gas. In the case of MoCl₅, it was understood that chlorine gas was generated by the disproportionation reaction of $MoCl_5$ to $MoCl_4$ and chlorine. Through this process, however, only a mixture of stage-1 and -2 structures of GICs was obtained, not pure stage-1. Recently, it was found that the addition of either $MoOCl_3$ or MoO_3 into $MoCl_5$ resulted in the formation of a stage-1 structure in a single phase [Mittal and Inagaki, 1998a, b], which was discussed to be due to the formation of a large amount of reactive chlorine gas from the reaction between $MoCl_5$ and either $MoOCl_3$ or MoO_3 [Mittal and Inagaki, 1999a, b].

A typical example of intercalation in an aqueous solution is the formation of H_2SO_4-GICs. The intercalation of H_2SO_4 proceeds by the addition of some oxidants, HNO_3, $KMnO_4$, etc., into concentrated sulfuric acid (chemical oxidation process). This process was understood to be comparable to electrochemical oxidation through the measurement of potential changes during these two processes, as shown in Fig. 6-18 [Iwashita and Inagaki, 1992]. The potential for each plateau, the onset potential, is definite for each stage structure irrespective of the oxidation processes, whether chemical oxidation by HNO_3, $KMnO_3$ or electrochemical oxidation. However, they were known from electrochemical studies to depend strongly on the concentration of sulfuric acid. In the case of chemical oxidation, the maximum potential reached depends strongly on the oxidant used and also the concentration of sulfuric acid. In Fig. 6-19, the values of onset potentials for each stage and the maximum saturated potential by HNO_3 are plotted against concentration of sulfuric acid. From this relation, it is understood that only the compounds with a stage number higher than 3 can be synthesized in sulfuric acid of 14 mol dm^{-3}, for example, because the maximum saturated potential of HNO_3 is slightly lower than the onset potential for stage-2 [Iwashita and Inagaki, 1989; Inagaki et al., 1990].

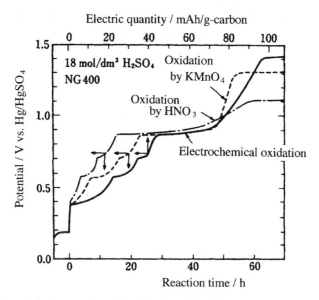

Figure 6-18   Potential changes of graphite with chemical oxidation by either $KMnO_4$ or HNO_3 and with electrochemical oxidation [Iwashita and Inagaki, 1992].

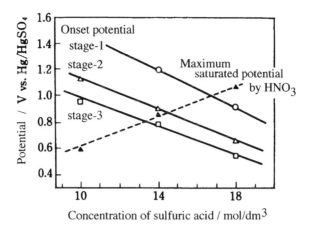

Figure 6-19 Relation between onset potentials for each stage structure and maximum saturated potential by HNO$_3$ oxidant as a function of concentration of sulfuric acid [Iwashita and Inagaki, 1989].

Intercalation of alkali metals in organic solutions has been found [Nomine and Bonnetain, 1969] and was recently studied extensively, reactions of which can be expressed by the following equations:

$$M + E + A \rightarrow M^+[E]_m + A^-, \tag{6-1}$$

$$C + A^- \rightarrow C^- + A, \tag{6-2}$$

$$C^- + M^+[E]_m \rightarrow C^-\{M^+[E]_m\}, \tag{6-3}$$

where M is alkali metals (Li, Na, K, etc.), E is ethers (tetrahydrofuran, 2-methylhydrofuran, diethoxyethane, etc.), A is an aromatic compound (anthracene, phenanthrane, etc.) and C is host carbon. Alkali metal dissolves into ether by the addition of an aromatic compound through electron transfer from alkali metal to aromatics, resulting in the formation of a complex of alkali metal ions solvated by the molecules of solvent ether (Eq. 6-1). Carbon has a slightly higher electron affinity than the aromatics used, and so charges on the aromatics in the solution are transferred gradually to solid carbon in the solution (Eq. 6-2). Then, the complexes $M^+[E]_m$ can intercalate into carbon by electrostatic attraction between these two (Eq. 6-3). The intercalation compounds thus formed are usually ternary ones because the intercalates are the complexes of alkali metal ions solvated by ether molecules, as shown in Fig. 6-8. Recently, however, binary compounds with alkali metals were found to be formed in some ether solutions, such as 2-dimethylhydrofuran (2-MeTHF) [Mizutani et al., 1996, 1997; Tanaike and Inagaki, 1997]. Whether ternary or binary compounds were formed was found to depend not only on the solvents used but also the host carbon materials [Tananike, 1998; Tanaike and Inagaki, 1998]. The intercalation of acceptor-type transition metal chlorides in organic solutions is also possible. In chloroform, two types of FeCl$_3$–CHCl$_3$-GICs with different structure in the intercalate layers were formed [Soneda and Inagaki, 1992]. In dimethylethane (DME), FeCl$_3$ intercalation was found to depend strongly on the graphitization degree of host carbons [Tanaike et al., 1998].

Table 6-3. Composition and stage number n of alkali metal-GICs synthesized under various conditions

Synthesis condition	Reaction with vapor			In organic solution		
Alkali metal	Normal pressure	Normal pressure with O_2	High pressure	In THF	In MeTHF	Electro-chemical
Li	LiC_6 $n=1$	–	LiC_2 $n=1$	Li-THF $n=1$	Li $n=1$	Li $n=1$
Na	NaC_{48} $n=6$	$NaC_6O_{0.6}$ $n=2$	$NaC_{2.6}$ $n=1$	Na-THF $n=1$	Na $n=10$	Na *
K	KC_8 $n=1$	$KC_{3.7}O_{0.07}$ $n=1$	KC_4 $n=1$	K-THF $n=1$	K $n=1$	K-DME $n=1$
Rb	RbC_8 $n=1$	RbC_4 $n=1$	$RbC_{4.5}$ $n=1$	Rb-THF $n=1$	Rb & Rb-MeTHF $n=1$	Rb-DME $n=1$
Cs	CsC_8 $n=1$	–	$CsC_{4.2}$ $n=1$	Cs-THF $n=1$	Cs-MeTHF $n=1$	Cs-DME $n=1$

* Stage number was not reported.

Sodium was known to have a very peculiar intercalation behavior, different from other alkali metals, Li, K, Cs and Rb, and special attention was paid to it [Tanaike, 1998]. In Table 6-3, the lowest stage structures of GICs obtained by different synthesis methods are tabulated as functions of alkali metal [Inagaki, 1995]. Direct contact of sodium vapor with graphite gives only a high stage structure, but if a trace of oxygen or sulfur and some organic molecules such as tetrahydrofuran (THF) are co-intercalated with sodium, a low stage, even stage-1, can be obtained easily. The co-intercalation of some organic molecules, e.g. THF, gives ICs with a stage-1 structure. In organic solutions such as 2-MeTHF, however, a binary compound is formed as mentioned above, but its stage number is as high as 10. Such a behavior has been explained by "misfitting" of the sodium lattice in the gallery with carbon hexagons [Dresselhaus and Dresselhaus, 1981].

Electrochemical intercalation of lithium into various carbon hosts, i.e. the formation of binary compounds with lithium in organic solutions such as ethylene carbonate, attracted attention as the fundamental electrode reaction in lithium rechargeable batteries and numerous studies have been performed and are ongoing worldwide [Dahn et al., 1994].

The intercalation reaction of fluorine to graphite was found to give two different compounds, covalent and ionic ones, depending on the reaction temperature [Nakajima et al., 1981, 1982]. At a reaction temperature of 350–600°C, it gives covalent compounds, such as $(CF)_n$, with puckered carbon layers (Fig. 6-12). At low temperatures such as 150–250°C in the presence of some metal fluorides such as LiF and CuF_2, however, fluorine is intercalated into the gallery of flat carbon layers by forming ionic interactions, where no covalent bond between carbon and fluorine occurs but electron transfer from carbon to fluorine occurs. The bonding nature between carbon and fluorine can be controlled by selecting proper conditions, providing new possibilities to create novel functions [Touhara and Okino, 2000].

Table 6-4. Possible applications of ICs and the origin of their functions

1. Highly conductive materials	
AsF_5, SbF_5, HNO_3, $CuCl_2$, $FeCl_3$, F_2, K–Bi, residual Br_2	Combination between host graphite and intercalates
2. Electrode materials in batteries	
Primary battery	
$(CF)_n$, $(C2F)_n$, graphite oxide, F_2, $CoCl_2$, TiF_4	Efficiency enhancement of function of intercalates
Secondary battery	
Li, H_2SO_4, $Ni(OH)_2$, $Mn(OH)_2$	Intercalation and deintercalation processes
Thermocell	
Br_2, HNO_3	Intercalation and deintercalation processes
3. Catalyzers for organic synthesis	
Li, K, K–Hg, K–$FeCl_3$, SbF_5, Br_2, H_2SO_4, HNO_3	Efficiency enhancement of function of intercalates
4. Materials for storage of gases and isotope-separation of hydrogen	
K, Cs, Rb	Creation of functional space in ICs
5. Others	
Exfoliation of graphite	
H_2SO_4, HNO_3, $FeCl_3$, K-THF, Na-THF	Intercalation and deintercalation processes
Thermal energy storage	
$MnCl_2$–NH_3	Intercalation and deintercalation processes
Electrochromic	
Li-DMSO	Intercalation and deintercalation processes

6.3. Functions Related to Applications

(a) Possible applications

With respect to the applications of ICs, their functions were discussed by dividing them into five categories, as summarized in Table 6-4 with representative intercalates and the main origin of each function [Inagaki, 1989a].

The high electrical conductivity of some ICs originates from the electronic structure change due to charge transfer between host carbon and intercalate, and therefore depends strongly on the combination between hosts and intercalates. High electrical conductivity, higher than that of metallic copper, was found on AsF_5- and SbF_5-GICs prepared from HOPG. Electrochemical and catalytic functions come from the intercalates themselves by enhancing their functions through intercalation into the carbon gallery. The recent development of lithium ion rechargeable batteries was principally based on the intercalation and deintercalation reactions into carbon at an anode. Space formed in ICs was found to play an important role in the function of adsorption and storage of gases by ICs. A large amount of hydrogen was found to be adsorbed into alkali metal-ICs with different selectivity of hydrogen isotopes. Different hydrocarbon gases were also adsorbed into alkali metal-ICs, some of which were confirmed to be oligomerized in the carbon gallery.

Other possibilities are the use of GICs for the production of exfoliated graphite, which is now an industrial process, passing through sulfuric acid-GICs, a proposal to use the intercalation/deintercalation reaction of ammonia into transition metal chloride-GICs as a reservoir for thermal energy, and the usage of the change in color to golden or blue by electrochemical intercalation of alkali metals as electrochromism.

(b) High conductivity function

AsF_5- and SbF_5-GICs were found in 1977 to have a high electrical conductivity, in the order of 10^8 Scm^{-1}, slightly higher than that of metallic copper [Vogel, 1977], which promoted research on various ICs. Some other ICs, such as $CuCl_2$- and $FeCl_3$-ICs, and also F-ICs with an ionic nature of bonding, have a rather high conductivity of 10^6 Scm^{-1}. In Fig. 6-20a and b, the electrical conductivity of ICs of various carbon fibers with AsF_5 and with other transition metal chlorides, respectively, is plotted against that of hosts [Endo et al., 1990].

In the case of AsF_5-ICs, their conductivity depends strongly on that of the hosts; in order to realize high conductivity in ICs the host needs to have high conductivity, in other words, high crystallinity (Fig. 6-20a). In the case of most transition metal chlorides, such as $FeCl_3$ and $CuCl_2$, however, the pristine carbon with the higher conductivity gives the IC with the higher conductivity, but it is almost 10 times larger than that of the host used, irrespective of the crystallinity of the host (Fig. 6-20b). The ICs with pentafluorides have higher conductivity, but have disadvantages in stability and safety in comparison with other chlorides.

Emphasizing the stability for aerospace applications, a residual compound of bromine prepared from carbon fibers was proposed, the electrical conductivity of which was reported to be able to reach 2×10^6 Scm^{-1} [Gaier and Jaworske, 1985]. For practical application as electrical conductors, not only their absolute value of electrical conductivity but also their light weight, stability and morphology have to be taken into account.

On alkali metal-GICs, KC_4 and CsC_4, synthesized under high pressure, super-conductivity was reported, the critical temperature of which was 5.5 and 6 K, respectively [Nalimova et al., 1992], although it was much lower than for ceramic superconductors.

(c) Electrochemical functions (electrode materials for batteries)

Electrode reactions which have been studied by using ICs are summarized in Table 6-5. Graphite fluoride $(CF)_n$, a covalent compound, has been used as an electrode material for lithium primary batteries (Eq. 6-5 in Table 6-5) since 1974. It was used successfully in primary batteries to give them many advantages: light weight, small size, relatively high voltage (about 3 V), high energy density (285 Wh kg^{-1}), high preservability and small self-discharge [Touhara et al., 1984b]. The compound $(C_2F)_n$ was found to give higher potential than $(CF)_n$ [Kita et al., 1979]. Another covalent compound, called graphite oxide, with a composition $C_4O(OH)$, was reported to give a high energy density (1200 Wh kg^{-1}), although the open-circuit voltage is slightly low (2.2 V) [Yazami and Touzain, 1983].

Recently, not only graphite but also various carbon materials were used as an anode for lithium rechargeable batteries, the principal electrode reaction being supposed to be intercalation and de-intercalation of lithium ions (Eq. 6-4). These lithium ion rechargeable batteries are already used in cellular phones and mobile computers. Industrial production has accelerated markedly since 1996, and has become greater than for other rechargeable batteries, Ni–Cd and Ni–metal hydride batteries in Japan, as shown in Fig. 6-21. Since the composition of stage-1 Li-GIC is LiC_6, the theoretical capacity of batteries using graphite as anode must be 372 Ah kg^{-1}. However, practical capacity values for most carbon anodes are lower than theoretical values, and depend strongly on the crystallinity of the host

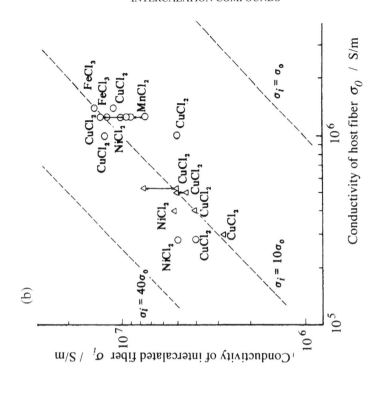

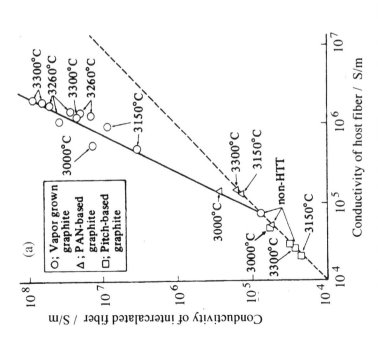

Figure 6-20 Electrical conductivity of ICs, σ_i, as a function of that of the host, σ_0: (a) intercalation of AsF_5 and (b) different metal chlorides into various carbon fibers [Endo et al., 1990].

Table 6-5. Electrochemical reactions related to IC electrodes

$nC + A^+ + e^- \leftrightarrow CnA$	$nC + Li^+ + e^- \leftrightarrow C_nLi$	(6-4)
$C_nA + B^+ + e^- \leftrightarrow nC + AB$	$(CF)_n + nLi^+ + e^- \leftrightarrow nC + LiF$	(6-5)
$nC + A^- \leftrightarrow C_nA + e^-$	$nC + H_2SO_4 + H_2O_4^- \leftrightarrow C_nH_2SO_4 \, nHSO_4 + e^-$	(6-6)
	$nC + ClO_4^- \leftrightarrow C_nClO_4 + e^-$	(6-7)
$C_nA + B^- \leftrightarrow C_nAB + e^-$	$C_nNi(OH)_2 + OH^- \leftrightarrow C_nNiOOH + e^- + H_2O$	(6-8)

carbon materials. In Fig. 6-22, the capacity of batteries with various carbon fibers as the anode are plotted against crystallite thickness $L_c(002)$, as a measure of crystallinity of the host carbon materials, showing a minimum at certain crystallinity [Endo et al., 1991, 1996]. A similar dependence of battery capacity on the crystallinity of anode carbon materials was reported on various carbon materials [Dahn et al., 1994]. In non-graphitizing carbons, the development of a crystalline structure is poor even after heat treatment at temperatures as high as 3000°C, and so low capacity was reported, which also revealed that heat treatment temperature cannot be a parameter to evaluate the anode capacity for carbon materials.

However, a very high capacity, much higher than the theoretical value, was found on some low-temperature-treated carbons. A value higher than 1000 Ah kg^{-1} was reported [Yata et al., 1994; Sato et al., 1994; Mabuchi et al., 1995]. The data on carbons derived from polyparaphenylene are plotted against $L_c(002)$ in Fig. 6-22, in order to show how the high-capacity value of low-temperature-treated carbon has been obtained. These carbon

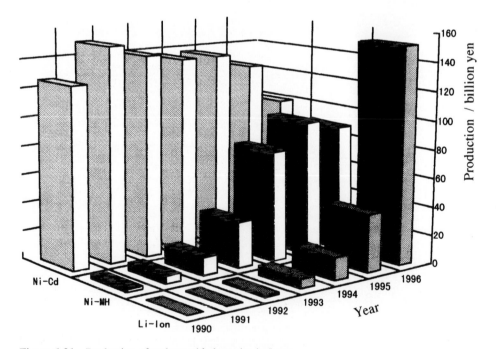

Figure 6-21 Production of rechargeable batteries in Japan.

materials usually show a very high irreversible capacity, which became a high barrier to their practical use as anodes. The reason why these low-temperature carbons can accept such a large amount of lithium ions is still controversial, including whether all of these lithium ions are intercalated into the carbon gallery. Three models in Fig. 6-23 have been proposed: in model (a) all lithium ions are registered to carbon layers and the theoretical composition is LiC_2 (capacity of 1120 Ah kg^{-1}) [Sato et al., 1994]; in model (b) lithium ions are located on both sides of single carbon layers, which are main structural units in these low-temperature carbons [Dahn, 1995]; and model (c) suggests a simultaneous occurrence of both intercalation into the gallery of small carbon layers and condensation at the pores (cavities) in these carbons [Mabuchi et al., 1995]. It is not yet known which model is correct.

A great prospect for these lithium rechargeable batteries is to extend their application to electric motor cars and electricity storage for its leveling is also anticipated. The recent development of lithium rechargeable batteries was reviewed by focusing on their research, industrial production and applications [Endo et al., 2000].

A variety of trials for battery applications has been reported, including intercalation/ deintercalation in HF solution using composite electrodes of natural graphite and polypropylene (Eq. 6-7) [Beck and Krohn, 1983], charging by chemical oxidation and electrochemical discharging using H_2SO_4 [Iwashita and Inagaki, 1994; Inagaki and Iwashita, 1994], transformation of $NiCl_2$ to $Ni(OH)_2$ in the graphite gallery to make

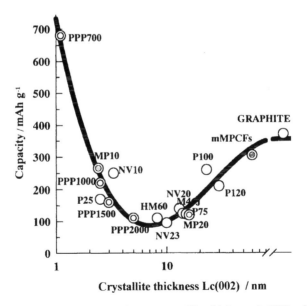

Figure 6-22 Dependence of anode capacity on crystallite thickness $L_c(002)$ of different carbon materials [Courtesy of Prof. Endo of Shinshu University.] P100, O120, HM60, P25, P75 and M46J are commercial mesophase-pitch-based carbon fibers; MP10 and MP20 are laboratory-made mesophase-pitch-based carbon fibers heat-treated at 1000 and 2000°C, respectively; NV10 and NV23 are vapor-grown carbon fibers heat-treated at 1000 and 2300°C, respectively; and PPP700, PPP1000, PPP1500 and PPP2000 are polyparaphenylene-based carbons with HTT of 700, 1000, 1500 and 2000°C, respectively.

effective usage of nickel hydroxide for Ni–Cd cells [Flandrois et al., 1981; Iwashita, 1992], and a simple secondary battery composed of carbon fiber anode and 30% KOH aqueous electrolyte [Otani et al., 1976].

A kind of concentration cell using either Br or HNO_3 which could convert the small temperature difference between two electrodes to electric power was proposed [Lalancette and Roussel, 1976; Inagaki et al., 1983b]. By using graphitized vapor-grown carbon fibers as the electrode, an open circuit voltage of 200 mV and short circuit current of 10 mA cm^{-2} were obtained from the temperature difference of about 80°C [Endo et al., 1980]. By using PAN-based carbon fibers and nitric acid, electric power was continuously obtained from waste hot water for 2 years without any maintenance action [Inagaki et al., 1990a].

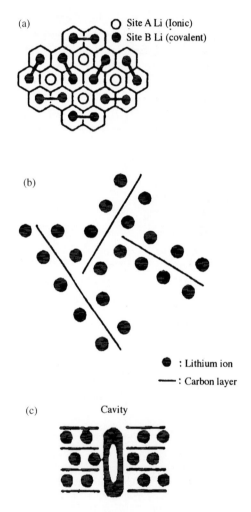

Figure 6-23 Models proposed to interpret a high anode capacity of low-temperature-treated carbon materials [Sato et al., 1994; Dahn, 1995; Mabuchi et al., 1994].

(d) Catalytic functions

Various ICs were tested for use in the synthesis process of different organic compounds, where the role of ICs may be divided into two cases, catalysts for accelerating the reactions and reagents supported in the carbon gallery [Gole, 1977; Setton, 1977].

In the former case, two cases have to be differentiated: ICs that act as catalysts in a strict meaning and do not show any change in structure, and ICs that help the formation of reaction intermediates but decompose during the reaction. Alkali metal-ICs were reported to be effective catalysts for the polymerization of various hydrocarbons such as ethylene and styrene, not only to increase the yield of polymerization but also to improve the selectivity. Published results of polymerization reactions of some hydrocarbons using alkali metal-ICs are summarized in Table 6-6 [Inagaki, 1988].

Examples of the latter are Br-ICs, which are reported to be a good reagent for bromination of hydrocarbons with a high selectivity and SbF_5-ICs for fluorination. $SbCl_2$-ICs were able to exchange bromine in organic molecules by chlorine.

(e) Gas adsorption and storage

It is known that K-ICs can react with hydrogen gas to form compounds with two different structures; one is $KC_8H_{2/3}$ formed from stage-1 KC_8 at room temperature and the other $KC_{24}(H_2)_2$ from stage-2 KC_{24} at around liquid nitrogen temperature [Lagrange and Herold, 1975]. Some characteristics of these two compounds are summarized in Table 6-7.

Table 6-6. Polymerization of hydrocarbons using alkali metal-ICs

Monomer	IC	Yield (wt%)	Composition of polymers	Conditions
Ethylene	KC_8	–	trans monoolefine	200°C, 68 bar, 21 h, iso-octene
Butadiene	KC_8	40–80	51%: 1–2, 49%: trans 1–4	30°C, 15 h, cyclohexane
	KC_{12}	51.3	90%: 1–4, 10%: 1–2	15°C, 100 h, cyclohexane
	KC_{12}	21.5	88%: 1–4, 12%: 1–2	15°C, 100 h, toluene
	KC_{24}	–	66%: 1–2, 35%: trans 1–4	75°C, 12 h, cyclohexane
Isoprene	KC_8	90	60%: 3–4, 5%: 1–2, 35%: trans 1–4	25°C, 12 h, n-heptane
	$LiCl_2$	80	44%: cis 1–4, 21%: 3–4, 35%: trans 1–4	15°C, 168 h, cyclohexane
	KC_{37}	95	43%: 1–4, 57%: 3–4	15°C, 76 h, cyclohexane
Methyl-methacrylate	$LiCl_2$	80	4%: iso, 63.6%: syndio, 32.4%: atactic	– 63°C, 48 h, DME
	$KC_8 + KC_{24}$	44	0.5%: iso, 50.5%: syndio, 49%: atactic	23°C, 4 h, DME

Table 6-7. Ternary K-H-GICs

	High-temperature form	Low-temperature form
Composition	$KC_8H_{2/3}$	$KC_{24}(H_2)_2$
Starting GIC	KC_8 (stage-1)	KC_{24} (stage-2)
Reaction temperature	Room temperature	77 K
Structure	Stage-2	Stage-2
	K–H–K triple layer	H_2 molecules between K
Formation energy	15–17 kcal mol^{-1}	2.2–2.9 kcal mol^{-1}
Isotope effect	Concentration of H	Concentration of D
Exchange	Formation of HD	No formation of HD

From the viewpoint of hydrogen storage, the latter compound, $KC_{24}(H_2)_2$ has certain advantages, including a small change in size accompanied by the adsorption/desorption of hydrogen, complete reversibility of adsorption and desorption of hydrogen, simple evacuation or heating for desorption, and only a small effect by contaminants such as oxygen. However, some disadvantages are also pointed out, such as the storage capacity being slightly smaller than a hydrogen storage alloy $LaNi_5H_6$ and the need for a low temperature, 77 K. The former compound, $KC_8H_{2/3}$, shows an isotope effect, with concentrated hydrogen rather than deutrium and tritium in the compound. It was shown experimentally that these K-GICs are effective column material for gas chromatography of hydrogen isotopes, only a 5 cm column of KC_{24} giving much better separation of p-H_2, o-H_2, HD, o-D_2 and p-D_2 around 77 K, as shown in Fig. 6-24, than a 5 m column of alumina [Terai, 1983].

Alkali metal-ICs were found to be able to adsorb different kinds of hydrocarbon gases [Takahashi, 1993]. In Table 6-8, the amount of hydrocarbons, the stage structure with I_c-value and the reversibility of the adsorption process are summarized on stage-2 CsC_{24}. In the case of n-alkanes, the adsorbed amount decreases with the increase in molecular size and all of these adsorption processes are reversible, easily desorbed by simple cryopumping. In contrast, some alkenes and alkynes containing double and triple C–C

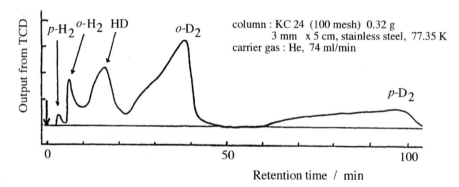

Figure 6-24   Gas chromatogram of hydrogen gas using a column of KC_{24}. [Courtesy of Prof. Terai of Tokyo University.]

Table 6-8. Adsorption of aliphatic hydrocarbons into CsC_{24} [Takahashi, 1993]

Hydro-carbon	Adsorption temperature	Composition of products				Reversibility of adsorption
		Stage-2		Stage-1		
		n	I_c	n	I_c	
n-Alkanes						
CH_4	90–181	1.2	0.986			Reversible
C_2H_6	194–252	0.9	0.993			Reversible
C_3H_8	252–287	0.69	1.001			Reversible
n-C_4H_{10}	275–326	0.55	1.010			Reversible
n-C_5H_{12}	300–327	0.54[a]	1.013–1.017			Reversible
n-C_6H_{14}	327–352	0.44[b]	1.016	0.9	0.694	Reversible
Alkenes and alkynes						
C_2H_4	194–323	1.1	1.006	2.1	0.685	Reversible below 200 K Irreversible above 273 K
C_3H_6	253–292	0.8	–			Reversible
C_2H_2	194–273	0.65	1.000			Irreversible

[a] With ~ 16% CsC_8 phase; [b] with ~ 26% CsC_8 phase.

bonds are irreversibly adsorbed into CsC_{24}. The isotherms of ethylene adsorption at different temperatures are shown in Fig. 6-25 [Pilliere et al., 1993]. Below 250 K, adsorption of ethylene C_2H_4 occurs in two steps, around 1.1 in C_2H_4/Cs at the first plateau and about 2.1 at the second. At 194 K, the adsorption is completely reversible, as in the case of *n*-alkanes. Above 273 K, however, only the first plateau is observed and the adsorption becomes irreversible. This irreversible adsorption was explained by the oligomerization of C_2H_4 molecules in the graphite gallery due to some catalytic action of Cs^+, on the basis of gas chromatography/mass spectrometry (GC/MS) analysis of toluene extracts from the compound and *in-situ* calorimetry of the adsorption process.

(f) Other functions

Abrupt heating of either GICs or their residue compounds induces a remarkable exfoliation of the host graphite perpendicular to its layers, mainly due to the rapid decomposition of intercalates to gaseous species. Flexible graphite sheets, which are prepared from exfoliated graphite by roll-forming without any binder, are currently in demand for use as gaskets, packing and thermal insulators at high temperatures [Fujii and Dohi, 1986]. A representative SEM micrograph of exfoliated graphite is shown in Fig. 6-26. The greatest advantage of these graphite sheets prepared from these exfoliated graphites is that any shapes and sizes are available, although their thickness is limited. In addition to the intrinsic properties of graphite, these graphite sheets show a high thermal resistivity, high anisotropy in thermal conductivity, stability to compressive strain, easy relaxation of induced stress, self-lubricity, etc.

In most industrial production of exfoliated graphite, residue compounds obtained from sulfuric acid-GICs by washing are used. In the process of this production, different

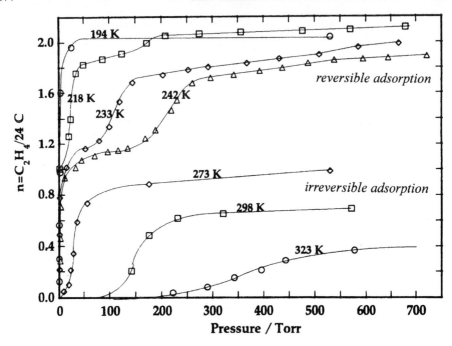

Figure 6-25 Adsorption isotherms of ethylene C_2H_4 into stage-2 Cs-ICs, CsC_{24}, at different temperatures [Takahashi, 1993].

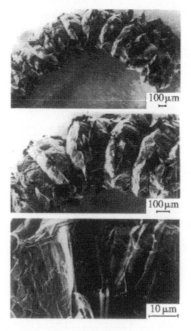

Figure 6-26 SEM micrographs of exfoliated graphite.

problems have to be overcome, related to the necessity of using concentrated sulfuric acid for the GIC production by either chemical oxidation or electrochemical oxidation, including the formation of poisonous oxide gases SO_x and NO_x during exfoliation, and the possibility of erosion of metals by a trace of sulfur remaining in the sheets through exfoliated graphite. Therefore, different intercalates, such as $FeCl_3$ and Na-THF, were tested to prepare exfoliated graphite [Berger and Maire, 1977; Inagaki et al., 1983a]. Recently, the synthesis of formic acid-GICs through electrochemical oxidation and their exfoliation by rapid heating were proposed to prepare exfoliated graphite without any sulfur [Kang et al., 1997].

The exfoliation behavior of carbon materials depends primarily on their texture. In Fig. 6-27a, the exfoliation process of vapor-grown carbon fibers heat-treated at high

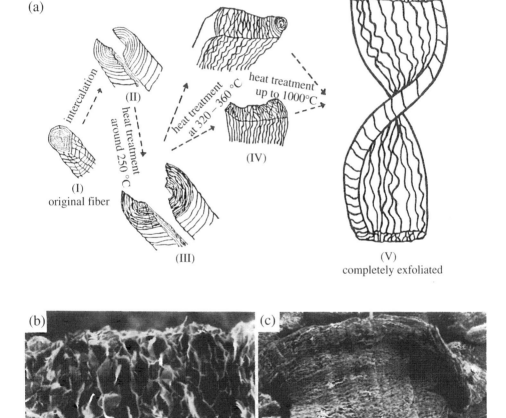

Figure 6-27 Exfoliation of vapor-grown carbon fiber heat-treated at high temperatures. (a) Schematic illustration of the exfoliation process; (b) SEM of the resultant carbon sheet [Yoshida et al., 1990].

temperatures is illustrated on the basis of SEM observations. Exfoliation occurs mostly in a radial direction of the carbon fiber having a concentric texture and, as a consequence, giving a sheet with hexagonal layers preferentially aligned perpendicular to its surface, as shown in Fig. 6-27b [Yoshida et al., 1990].

Recently, these exfoliated graphites were found to sorb a large amount of heavy oil very quickly, more than 80 g of heavy oil per 1 g of exfoliated graphite within 1 min, which seemed to be promising for the recovery of heavy oil dispersed in water, as explained before (Chapter 5) [Toyoda et al., 1998a, b, 1999]. The experimental results on sorption of different heavy oils into exfoliated graphite and their recovery were recently reviewed [Toyoda and Inagaki, 2000] and a new research project was started by the New Energy and Industrial Technology Development Organization (NEDO) in Japan.

A system for the storage of thermal energy has been proposed [Touzain et al., 1983], in which thermal energy was stored during deintercalation of ammonia molecules from ternary GICs with metal chloride and NH_3, and released by a reverse reaction, i.e. formation of ternary GICs by the intercalation of NH_3 into binary metal chloride-GICs. A thermal capacity of about 200 Wh kg^{-1} of GIC was reported.

A reversible change in color, from black in graphite to blue in Li-GIC, due to intercalation and deintercalation of Li in the organic electrolyte was tested for use in electrochromic devices [Pfluger et al., 1979; Yoshino and Ueno, 1989].

CHAPTER 7

Carbon Composites

7.1. Carbon Materials as Composites

Many carbon materials which have been produced in industry are composites in a broad sense. These composites may be classified on the bases of constituent materials, their shape (dimensionality) and their size of mixing in order to achieve a combination of functions, as shown in Fig. 7-1.

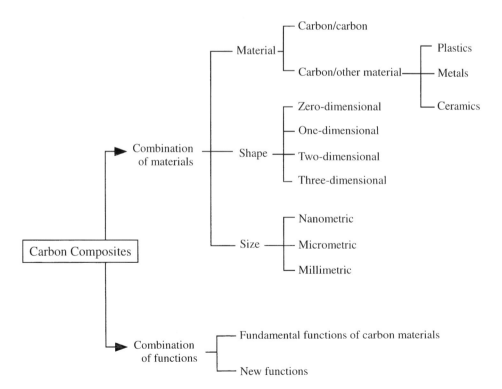

Figure 7-1   Classification of carbon composites.

According to this classification of composites, intercalation compounds (Chapter 6) are composites on a nanometer scale between the host carbon and intercalates of other materials, isotropic high-density graphites (Chapter 3) are composites on a micrometer scale between particles of filler coke and carbon derived from binder pitch (binder coke), as well as similar carbon materials composed of filler and binder cokes such as graphite electrodes, various porous carbons (Chapter 5) may be composites in micrometer and/or nanometer scale between carbon matrix and pores, etc. Among monolithic carbon materials, carbon fibers (Chapter 4) are the most important component with a one-dimensional shape of various composites with different matrices, such as resins, carbons and ceramics. Most precursors used to produce glass-like carbons (Chapter 3), such as phenol resin, are an important component as a matrix for carbon/carbon composites.

In the present chapter, the description focuses mainly on the composites of carbon fibers with other materials, i.e. carbon fiber reinforced composites, because most other composites according to the concept in Fig. 7-1 have already been explained in previous chapters. Here, carbon fiber reinforced plastics which developed their production in industry and their applications are described in brief by emphasizing their new applications, and then recently developed composites, i.e. carbon fiber reinforced concrete, ceramics and carbon, are described. All three can be abbreviated to CFRC, but to avoid confusion and according to convention, CFRC, CFRCer and C/C, respectively, will be used in this chapter.

7.2. Carbon Fiber Reinforced Plastics

Carbon fiber reinforced plastics (CFRP) are very important structural materials in different fields of industry, not only for aerospace but also for sport and leisure goods. In Table 7-1,

Table 7-1. Application of carbon fiber reinforced plastics (CFRP) and their merits

Applications	Merits
Sporting goods	
Golf clubs, rackets for tennis and badminton	Light weight, fashionable
Fishing rods	Light weight, fashionable
Skis	Light weight, fashionable
Sailboat spars	Light weight, rigidity, anticorrosion
Aerospace	
Primary aircraft structure, wing, fuselage	Light weight, high fatigue strength
Satellite bodies	Light weight, high rigidity, size stability, high thermal diffusivity
Automobiles	
Shafts	Light weight, high speed rotation
Wheel, bonnet	Light weight, high fatigue resistance
Parts for engine	Light weight (reduction of inertia)
Industry	
Robots	Light weight, less inertia
Chemical plants	Anticorrosion, less electrostatic charging
Medical instruments	Transparency for X-ray, high rigidity
Rotary blades	Light weight, high speed rotation

the applications of CFRP and its merits are summarized. In addition to the high strength and modulus of carbon fibers, different merits are obtained by using these CFRP. Practical problems relating to their production and applications have been discussed from different engineering viewpoints in the literature [Plastics and Rubber Institute, 1986].

A large amount of carbon fibers, mostly PAN-based ones, are used in rackets for badminton and tennis, golf clubs and fishing rods. At first, the use of carbon fibers gave a feeling of quality to customers, but it is now common in many sporting goods. The application to airplanes was developed quickly in military fields, but now even business jets are produced by fully using CFRP to reduce their weight. The application to cars is not so great, mainly because of the high cost of carbon fibers, even though CFRP can give a low weight to the body. However, low weight car bodies are so strongly required, in order to save energy and reduce waste gases, that the use of CFRP will probably become essential. CFRP is gradually starting to be used in parts of robots and other industrial equipment and is expected to be expanded in the near future.

Prepreg sheets of carbon fibers have been used for reinforcing the piers of highways and in chimneys. In Japan, after the large earthquake in the Kobe area (January 17, 1995) and also because of a large increase in traffic in the Tokyo area, the urgent reinforcement of piers of highways and pillars in buildings was pointed out. A typical example of reinforcement is shown schematically in Fig. 7-2. The tapes of carbon fiber prepregs were first glued onto the surface of either the pier or chimney along its axis, after finishing the surface of the concrete, and then either the same prepreg tapes or carbon fiber strands of 1200–2000 fibers were wound up perpendicular to the pier or chimney axis. In the case of a chimney which has been eroded by highly acidic exhausts, the advantages of light weight and easy repair from the outside of the chimney were recognized and about 20 cases of repair have been done in Japan. The tallest chimney repaired was 100 m high. Figure 7-3 shows a snapshot of the practical winding process of the carbon fiber strand onto a pier using remote-controlled equipment.

Most highway piers have a square cross-section and special attention was required at the corners so as not to break the carbon fibers. Many new technologies have been introduced for these repairs, including a remote-controlled system to check the corrosion of large-size concrete constructions, and a technique passing electrical current in order to harden the plastic matrix after forming.

A recent marked development in reinforcement of the glued laminated timbers by placing a thin sheet of carbon fiber prepreg was reported [Ogawa, 2000], in order to apply cryptomeria (Japanese cedar), which has been planted in large areas in Japan, for the construction of houses and bridges.

7.3. Carbon Fiber Reinforced Concrete

(a) Application to constructions

Reinforcement of concrete by using GP-grade chopped carbon fibers led to prominent success and was applied to various constructions [Inagaki, 1991]. The first application of carbon fiber reinforced concrete (CFRC) was in the Al Shaheed monument in Baghdad, Iraq (Fig. 7-4a). The steel skeleton and foundation of Al Shaheed monument, consisting of twin domes with a height of about 40 m and bottom diameter of about 45 m, were constructed according to the original plan to cover the domes with gold-plated copper plates. However, the replacement of copper plates by porcelain tiles of a Turkish blue color was decided after finishing the steel skeleton. Therefore, the cladding tile panels needed to have a weight of less than 60 kg cm^{-2}. In addition to light weight, enough mechanical strength and durability for the severe weather conditions in Baghdad, high temperature

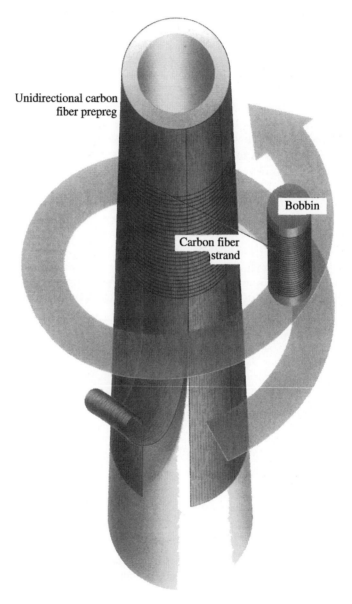

Figure 7-2   Reinforcement of a concrete pier using tapes of carbon fiber prepregs and carbon fiber strands. [Courtesy of Mitsubishi Rayon Co. Ltd.]

Figure 7-3 A snapshot of the winding process of a carbon fiber strand by remote-controlled equipment. [Courtesy of Mitsubishi Rayon Co. Ltd.]

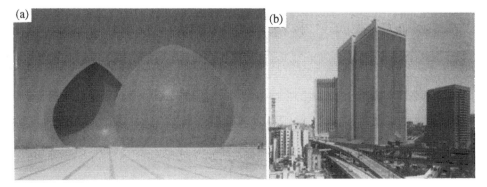

Figure 7-4 Al Shaheed monument (a) and Ark Tower building (b), in which carbon fiber reinforced concretes (CFRC) were used. [Courtesy of Dr Akihama of Kajima Construction Co. Ltd.]

and low humidity in summer and below zero temperatures in winter, were required. The light-weight CFRC was selected as the only one which satisfied all of these quality requirements [Akihama et al., 1984]. In this dome, a large number of CFRC panels (about 10,000 m^2) with a very delicate curvature was used, all of which were produced in Japan and, after autoclave curing, transported to Iraq. This success led to the application of this CFRC to the Ark Tower office building in Tokyo, with 37 stories (Fig. 7-4b), as curtain walls (32,000 m^2), which led to a marked reduction in wall weight, easy transportation by a small lift and, as a consequence, shortening of the construction period [Akihama, 1988]. These prominent successes led to further application of CFRCs in different components of different constructions [Nakagawa, 1996].

CFRCs have been used in different parts of buildings, curtain walls, waterproof roofs, floors with chemical stability and low level of noise, etc., and now they are used not only in buildings but also in the construction of bridges, tunnels, etc.

CFRCs have been found to be effective for electromagnetic shielding in modern buildings (sometimes called intelligent buildings), in order to block out electrical and/or magnetic noises from the outside and maintain the steady operation of the building's functions by computers. The electrical shielding effectiveness of CFRCs using carbon fiber non-woven mats is shown in Fig. 7-5 [Kojima et al., 1990]. By the addition of only one sheet of carbon fiber mat (corresponding to a carbon fiber content of 0.5 vol%), a shielding effect of 15–45 dB is obtained over the whole range of frequency of the electrical field. An increase in the number of mats increases the shielding effect. Recently, it was successfully used in piles with a demagnetization function in the Yokohama Bay area in Japan. They have been also used as a main component of free-access floor panels due to the following advantages: light weight, high strength, high dimensional stability, good heat insulation and low walking noise level, in comparison with the aluminium die-cast floor commonly used.

An example of a new application of carbon fibers is shown schematically in Fig. 7-6. It is a thin mortar plate reinforced by carbon fiber fabrics and chopped ones for use in frames for concrete placing. Many advantages have been pointed out, such as easy forming by bending at the site of construction, as shown in the figure, easy transportation and storage because of its light weight, thin thickness, chemical and mechanical stability.

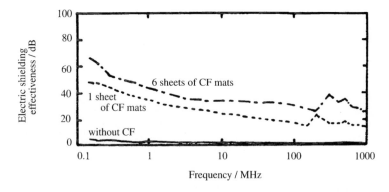

Figure 7-5 Electrical shielding effectiveness on CFRCs [Kojima et al., 1990].

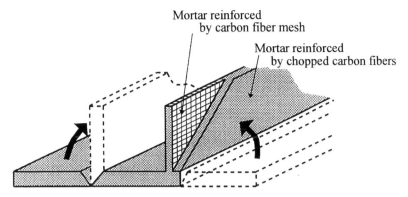

Mortar reinforced
by carbon fiber mesh

Mortar reinforced
by chopped carbon fibers

Figure 7-6   Frame of carbon fiber reinforced mortar for concrete placing. [Courtesy of Toho Rayon Co. Ltd.]

It was also useful for saving labor because it did not need to be taken off. CFRCs were also used for tunneling with automatic drilling tools to strengthen the tunnel walls. It was reported to be used successfully at a depth of about 73 m.

Two novel methods for reinforcing concrete by using carbon fibers have been developed, one being the use of a rod of carbon fiber reinforced plastics instead of steel wires in concrete, and the other the use of prepreg sheets (tapes) of unidirectionally aligned carbon fibers with a polymer matrix, the latter having been described in the previous section.

Rods consisting of a few strands of carbon fiber reinforced plastics have been tested for reinforcing a small concrete bridge, which is directly exposed to the erosive sea wind in Ishikawa Prefecture facing the Japan Sea [Katawaki and Nishizaki, 1988]. Prior to this, it was necessary to reconstruct the bridge at least once a year because of rapid erosion of the steel wires in the concrete. By using rods of carbon fibers, no defects in the bridge were observed even after 6 years. It has been used successfully in a pedestrian bridge crossing a main road in Funabashi City, Chiba Prefecture, since 1989.

(b) Reinforcement effect of carbon fibers

Many fundamental and engineering studies on CFRCs have been carried out using not only chopped PAN-based and pitch-based carbon fibers but also mats (papers) and fabric of continuous carbon fibers [Inagaki, 1991].

The addition of only 2 vol% of GP-grade isotropic-pitch-based carbon fibers to the cement mortar doubles the tensile strength, as shown by stress–strain curves for CFRCs with different volume fractions of carbon fibers in Fig. 7-7. Not only a marked increase in tensile strength but also a change in fracture mode from a brittle to a ductile one with the addition of carbon fibers is clearly seen. Figure 7-8 shows the dependence of initial crack strength, which is defined by the first break on the stress–strain curve, and ultimate fractured strength in tension on carbon fiber content under two different curing conditions, curing in air, i.e. keeping in a room at 20°C with 65% humidity for seven days, and autoclaving, i.e. curing in an autoclave at 180°C under a pressure of 1 MPa of steam for 5 h and then holding in a room maintained at 20°C with 65% humidity. A prominent

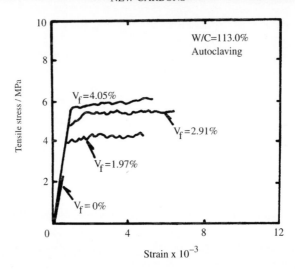

Figure 7-7 Stress–strain curves for CFRCs with different volume fraction of carbon fibers [Akihama et al., 1984].

increase in tensile strength with the increase in carbon fiber content and also by autoclaving is seen. The high resistance of carbon fibers to alkaline solutions makes

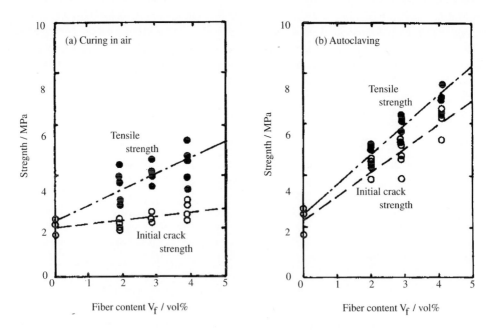

Figure 7-8 Dependence of strengths in tension on carbon fiber content under two different curing conditions [Akihama et al., 1984].

autoclave treatment of CFRC possible. Various durability tests of the CFRCs were performed before applying them to the Al Shaheed domes. Keeping in hot water at 75°C and repeated cycles between freezing at –18°C and thawing at 10°C were confirmed not to give any changes in mechanical properties and dimension. These results showed satisfactory durability of the CFRC for severe changes in environmental conditions, e.g. transportation from Japan to Iraq through the Indian Ocean.

The fracture behavior under tensile and bending stresses was studied in detail on concrete reinforced by unidirectionally aligned continuous GP-grade pitch-based carbon fibers [Akihama et al., 1984]. Typical tensile stress–strain curves along the fiber direction are shown in Fig. 7-9. Cracks occur between points A and B, represented on the CFRC with 4.72 vol% carbon fibers in Fig. 7-9, and most of them run perpendicular to the loading direction, i.e. perpendicular to the fiber direction, as shown in Fig. 7-10. With the increase in strain from point A, the number of cracks gradually increases and so their spacing decreases. Beyond point B, no additional formation of cracks is observed. In this region, all of the load seems to be supported by the carbon fibers, and so the slope of the stress–strain curve between points B and C is roughly the same as that calculated from Young's modulus and the content of carbon fibers. In Fig. 7-11, the dependence of ultimate tensile strength on fiber content is shown for two water/cement ratios. Above a certain fraction of fibers, the tensile strength of the CFRCs increases rapidly with increasing fiber fraction.

Non-woven mats of carbon fibers were also used to reinforce concrete from Portland and alumina cements [Kojima et al., 1989, 1990], because of their advantage of being easily placed where high stress and electromagnetic shielding are expected. Carbon fiber

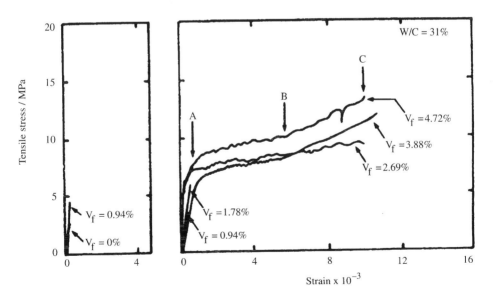

Figure 7-9 Stress–strain curves of concrete plates reinforced by unidirectionally aligned continuous isotropic-pitch-based carbon fibers with different content V_f under tensile stress [Akihama et al., 1984].

mats were laid up by hand in a cement paste without any aggregate and then cured in water at 40°C for 4–13 days. In the use of carbon fiber mats, the particle size of the cement was important for the penetration of cement paste into the voids in the mats, and consequently for achieving high strength. In Fig. 7-12, the bending strength of the CFRCs with alumina and Portland cements is plotted as a function of the number of carbon fiber mats laid up; the bending strength of CFRC reaches 72 MPa by lay-up of 25 sheets, which is about 18 times larger than the value without carbon fibers.

By using carbon fiber mats and alumina cement, a very thin plate of CFRC can be prepared. Figure 7-13 shows a thin plate of CFRC (1 mm thick), prepared from PAN-based carbon fiber mats and alumina cement, and how it can be deflected, showing about 33 mm deflection, which would not be possible without carbon fibers.

The surface treatment of carbon fibers by a low-temperature oxygen plasma was found to help the wetting of the carbon fiber surfaces with cement paste and consequently to increase the strength of the resultant CFRCs [Kojima et al., 1989]. In Fig. 7-14, the load–deflection curves in three-point bending test are shown on the CFRCs with 12 sheets of carbon fiber mats which were treated in advance by oxygen plasma for 20 min with

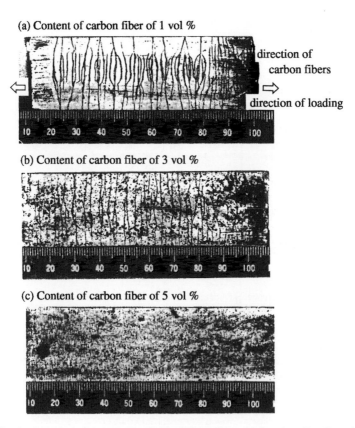

Figure 7-10 Appearance of a concrete plate reinforced by unidirectionally aligned continuous isotropic-pitch-based carbon fibers under tensile stress [Akihama et al., 1984].

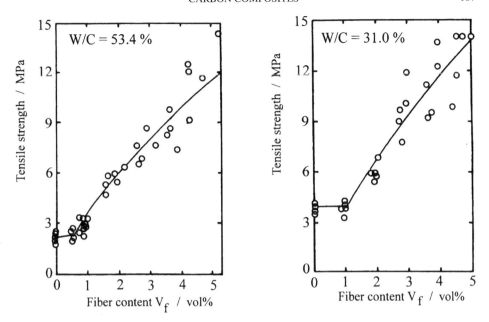

Figure 7-11 Dependence of ultimate tensile strength on fiber content for two different water/cement ratios [Akihama et al., 1984].

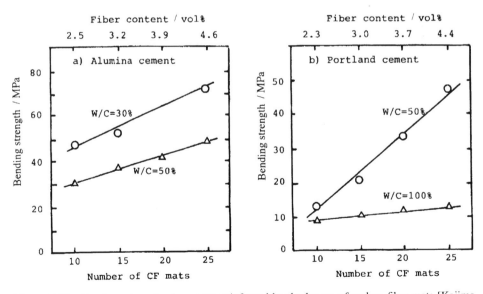

Figure 7-12 Bending strength of concrete reinforced by the lay-up of carbon fiber mats [Kojima et al., 1989].

different powers. Plasma treatment with a power of only 50 W makes the strength of CFRC twice as large as that without plasma treatment. The strength becomes more than 30 times greater than in concrete without carbon fibers. Irradiation with a higher power may introduce defects in carbon fibers and gives lower performance.

CFRCs have an additional advantage, i.e. the possibility of detecting fractures, even localized fractures, from the change in electrical resistivity of the filler carbon fibers [Muto et al., 1992]. This principle has already been applied to practical buildings, not only to detect partial fracture due to fatigue but also to monitor intentional breaks.

7.4. Carbon/Ceramics Composites

(a) Carbon fiber reinforced ceramics (CFRCer)

The main purpose of making composites of ceramics with carbon fibers is to give them toughness and so to avoid a catastrophic breaking of ceramic parts. Figure 7-15 shows an example of fracture behavior observed on a CFRCer, carbon fiber/Si_3N_4, where carbon fiber bundles are unidirectionally aligned [Mitsuoka et al., 1994]. With increasing load, deflection initially increases gradually and then increases without an increase in load. Compared to the fact that monolithic Si_3N_4 shows a linear increase in deflection with increasing load and then suddenly breaks (catastrophic breaking), the fracture behavior of

Figure 7-13   Large deflection of a thin plate of alumina cement reinforced by PAN-based carbon fiber mats. [Courtesy of Prof. Kojima of Gunma National College of Technology.]

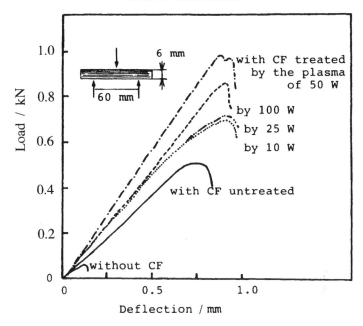

Figure 7-14 Load–deflection curves of CFRCs with or without oxygen plasma treatment on carbon fibers [Kojima et al., 1989].

the composite becomes ductile, i.e. cracks propagate gradually in the composite and no catastrophic breaking occurs. In Fig. 7-15, the electrical resistivity change ΔR is also plotted against deflection, showing a good correspondence of ΔR to change in load–deflection curve, i.e. crack propagation behavior. Even saw-tooth-like small changes in the load–deflection curve, which are supposed to be partial breaking of the carbon fibers, show good correspondence to ΔR. This result indicates a possibility of detecting partial fracture in this composite, which may give an additional advantage in using these

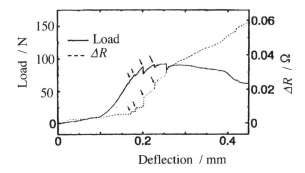

Figure 7-15 Load–deflection and electrical resistivity change ΔR–deflection curves for a composite of Si_3N_4 with 40 vol% carbon fibers. [Courtesy of Dr Matsuo of NGK Spark Plug Co. Ltd.]

composites as structural components. The same technique has been already used in practice in the case of CFRCs, as described above.

To produce a composite from a combination of carbon and ceramics, chemical and physical compatibility between these two components are the most important factors to be controlled. Most ceramics crystallize at temperatures as high as 1600–1800°C and, as a consequence, the composites have to have a thermal history up to these temperatures. At such high temperatures, chemical interactions between carbon and ceramics are very important to achieve the required properties. To prepare a composite of carbon fiber and mullite ($3Al_2O_3 \cdot 2SiO_2$), for example, a sol was firstly prepared from alkoxides of aluminium and silicon, and then chopped carbon fibers were dispersed in this sol, followed by gelation at relatively low temperatures such as 100°C. In order to crystallize the gel to the mullite structure, it has to be heated at about 1800°C, where chemical reactions of carbon fibers with decomposition products from alkoxides or oxides are expected, most of which have to be avoided because of their destructive effect on the carbon fiber structure. Therefore, not only carbon fibers but also the raw materials for mullite and their preparation procedure have to be selected.

Physical compatibility between the carbon fibers and ceramic matrix also has to be taken into consideration, which is caused by the stress at the boundary between these two components due to a difference in their thermal expansion during the cooling process from the high preparation temperatures. To prepare the composites, the formation of cracks due to these stresses has to be avoided, but the accumulation of a certain amount of these stresses may help the reinforcement of the composites. When the thermal expansion coefficient of matrix ceramics, α_M, is larger than that of filler carbon, α_F, a stress σ formed at their boundary is expressed as follows:

$$\sigma = E\,(\alpha_M - \alpha_F)\,\Delta T/(1-n),$$

where E is Young's modulus of the composite, n the Poisson ratio and ΔT the temperature difference from the preparation to room temperature. A large stress, larger than can be stored in the structure, may cause cracking. Therefore, combinations between carbon and ceramics with a small difference in thermal expansion coefficient, in other words, ceramics with a similar thermal expansion coefficient to carbon, are preferable.

(b) Metal carbide/carbon composites

The preparation of the composites of carbon with metal carbides, mainly boron carbide, has been extensively studied [Kobayashi et al., 1988]. Two processes were employed, hot-pressing using calcined coke (Process A) and pressureless sintering using raw coke (Process B), as shown in Fig. 7-16.

In Fig. 7-17a and b, the relative density and bending strength of the composites B_4C/C are plotted against the mixing ratio of boron carbide B_4C, respectively, as a function of the hot-pressing temperature. Densification and strengthening are observed above 2000°C, which is assumed to be caused by the diffusion of boron atoms into carbon particles. The composites thus prepared showed high resistance to oxidation in air at 800°C and also in sulfuric acid with 10% nitric acid, no change in weight except for a slight increase in surface roughness being detected.

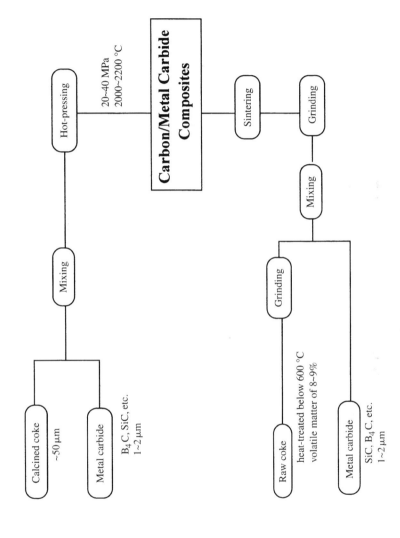

Figure 7-16 Preparation procedures for metal carbide/carbon composites [Kobayashi et al., 1988].

At the very beginning of oxidation, the formation of a thin layer of boron oxide on the surface of the composites was recognized and was supposed to act as a protective layer for further oxidation. The composites were also resistant in molten metals, such as aluminium and copper. In Fig. 7-18, the decrease in thickness of a blade of B_4C/C composite with number of heating cycles in molten aluminium is compared with conventional polycrystalline graphite blades. The improvement in the corrosion resistance of the composites is pronounced.

The addition of a third component of metal carbide, such as NbC, TaC or TiC, was effective in increasing the strength and thermal shock resistance of the composites. They have a much higher strength than graphite materials up to 1700°C, as shown in Fig. 7-19. In these composites, the added metal carbides were found to change to borides such as NbB, TiB_2 and TaB_2, after hot-pressing.

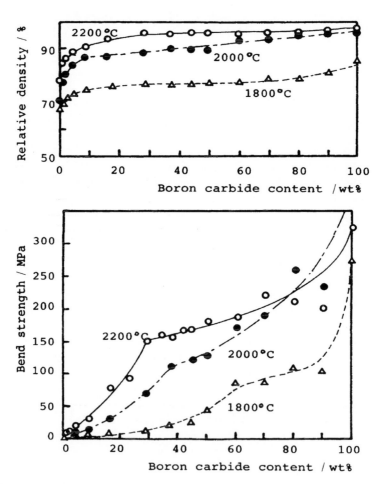

Figure 7-17   Dependence of relative density (a) and bending strength (b) of the composites on boron carbide content [Kobayashi et al., 1988].

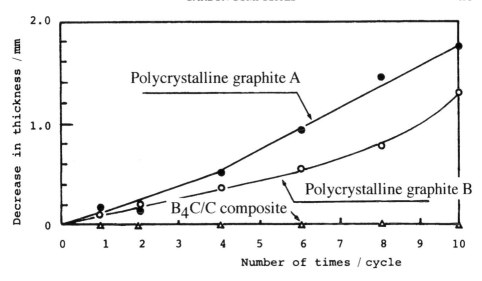

Figure 7-18 Decrease in thickness of blades of different materials with heating cycle in molten aluminium. B_4C/C composite was prepared from a mixture of raw coke with 30% B_4C (Process B in Fig. 7-16) [Kobayashi et al., 1988].

When raw coke powder, which was manufactured at about 500°C and contained about 8.5 wt% volatile matters, was ground in a mortar for a long time, it was possible to sinter under atmospheric pressure (pressureless sintering; Process B in Fig. 7-16) [Ogawa et al., 1981]. This coke powder was mixed with carbide powder, such as B_4C or SiC, and ground again in a mortar. The mixture formed small balls with a size of 1–5 μm, in which the carbide particles were well dispersed (Fig. 7-20), and sintered into the composites at 1000–2400°C without any pressure, in which small carbide particles were dispersed uniformly in a carbon matrix. Some properties of the B_4C/C composites thus prepared are shown as a function of B_4C content in Fig. 7-21. The addition of B_4C is again very effective in increasing the bending strength. Boron was substitutionally incorporated into carbon hexagonal networks up to about 3 at% [Lowell, 1967; Hishiyama, 1998]. Associated with this solid solution of boron, the crystallinity of the carbon matrix was found to be improved [Miyazaki et al., 1977; Sogabe et al., 1996, 1997] and also to increase the oxidation resistance of carbon materials. Composites with other carbides were also studied [Ogawa et al., 1988; Kobayashi et al., 1988].

Pressureless sintering is advantageous for the preparation of large-sized and complex-shaped products. In addition, these composites are machinable; in particular, electric discharge machining is easily applied. Some industrial products of these pressureless sintered $B_4C/SiC/C$ composites are now on the market.

(c) Coating of ceramic films on carbon materials

Ceramics coating has mostly been carried out in order to improve the oxidation resistance of carbon materials at high temperatures. The coating has been performed by either impregnation, CVD or dipping into precursor sols [McKee, 1997].

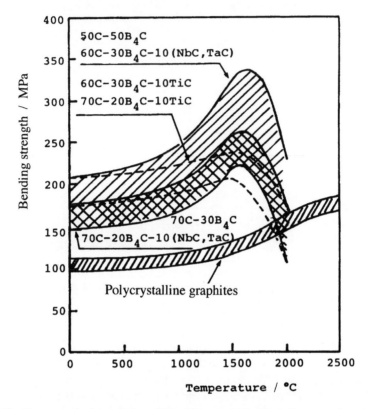

Figure 7-19 Temperature dependence of bending strength of various metal carbide/carbon composites compared with polycrystalline graphite. Here, 60C–30B$_4$C–10TiC, for example, stands for the composite prepared from a mixture of 60% raw coke, 30% B$_4$C and 10% TiC by Process B in Fig. 7-16 [Kobayashi et al., 1988].

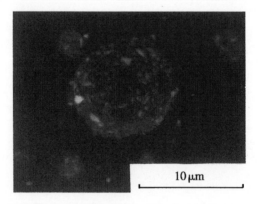

Figure 7-20 Optical micrograph of carbon/boron carbide composite showing a uniform distribution of small particles of B$_4$C. [Courtesy of Dr Ogawa of Government Research Institute of Kyushu.]

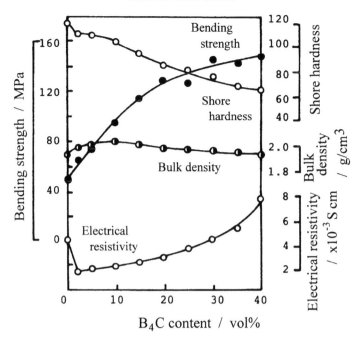

Figure 7-21 Properties of B₄C/C composite as a function of B₄C content [Kobayashi et al., 1988].

Boron oxide was used to coat carbon materials by different techniques, of which the principal role was to form a glass layer on the surface and avoid contact between oxygen gas and carbon at high temperatures [Ehrburger et al., 1986], as in the case of boron carbide/carbon composites described in the last section. Antioxidation was performed effectively in the temperature range of 600–1000°C, but no oxidation resistance was obtained at low temperatures, below 600°C or high temperatures above 1000°C, because of the lack of formation of a glass layer and the vaporization of boron oxide, respectively. In order to attain a high performance of the boron oxide coating, the addition of other oxides, such as silica, which modify the property of the glass phase, was tried [Kobayashi et al., 1988]. Coating of some oxides for use as frits for porcelains was also reported [Tanabe et al., 1987].

The deposition of silicon carbide SiC on the surface of carbon materials was tried by different methods and its effectiveness reported for oxidation resistance [McKee, 1986, 1988; Nixon and Cawley, 1992]. In this case, the outermost layer changes to silica at the beginning of oxidation to form glass phase, as in the case of boron carbide. Recently, a simple process to form a SiC concentration gradient in the carbon matrix was developed, which consisted of dipping the carbon materials into molten silicon at 1450°C by selecting an appropriate ratio of silicon to the physical surface area of the carbon material [Yamamoto et al., 1993].

A coupling of the formation of a SiC concentration gradient in carbon materials with overcoating of either zircon (ZrSiO₄) or mullite (3Al₂O₃ 2SiO₂) on their surfaces proved to be effective in having a good oxidation resistance at high temperatures as 1400°C [Yamamoto et al., 1994, 1995]. The coating of either zircon or mullite was performed by dipping the substrates with a SiC concentration gradient into the precursor solutions, which were prepared from $Zr(OC_4H_9)_4$, $Si(OC_2H_5)_4$ and $Al(NO_3)_3$ by selecting appropriate procedures. The sol films on the substrates thus prepared were kept at room temperature to gelation. The gel films on the substrates were heated at high temperatures in Ar flow to crystallize to either zircon or mullite. The final heat treatment temperature for zircon films was 1000°C and that for mullite was 1300°C. In Fig. 7-22a and b, the changes in weight loss per unit physical area at 1000 and 1400°C, respectively, with oxidation time in an air flow of 200 ml min^{-1}, are compared on the as-received carbon/carbon composite, that with a SiC gradient, and those coated with zircon film of 0.5 and 1.5 μm thickness [Yamamoto et al., 1995]. On the carbon/carbon composite with low bulk density, the formation of a SiC gradient in the substrate was not found to be effective at either 1000 or 1400°C. For an improvement in oxidation resistance of carbon materials at high temperatures such as 1400°C, a coupled use of SiC concentration gradient and overcoating of zircon film with a thickness of 1.5 μm was clearly shown to be effective. The oxidation loss was negligibly small at 1000°C and was only few milligrams per square centimeter of physical surface area of carbon materials at 1400°C after 30 h. From the XRD measurement, the surface oxidized at 1000°C for 30 h gave only diffraction peaks of zircon. By oxidation at 1400°C, however, small diffraction peaks of tetragonal-ZrO_2 and α-SiO_2 were detected in addition to those of the main phases of zircon and carbon. These additional phases were assumed to be formed by the phase separation of zircon at the contact with substrate carbon at high temperatures.

Coating of mullite film with a thickness of 1.0 μm, coupled with the SiC gradient in the matrix, was found to be effective to avoid the oxidation of carbon materials [Yamamoto et al., 1994]. On the mullite-coated specimens, rapid heating up to 1400°C and quenching down to room temperature were repeated in air flow, and weight loss measured at room temperature. The results are shown in Fig. 7-23. No weight loss was detected even after five repetitions of rapid heating and quenching. The last quenching was done by immersing the specimen into liquid nitrogen temperature (–196°C). From XRD, only the mullite phase was observed on all quenched specimens, indicating no decomposition of the mullite phase. Under SEM, no cracks were observed on the specimen surfaces after quenching to room temperature and only a few after quenching to liquid nitrogen temperature.

There has been a proposal [Strife and Sheehan, 1988] on the fundamental concept of obtaining sufficient oxidation resistance of carbon materials to form multilayers consisting of:

oxide/silica glass/refractory oxide/refractory carbide/carbon.

However, it is very difficult in practice to find a good combination to satisfy the physical compatibility between these materials, i.e. matching their thermal expansion coefficients at three boundaries between two components of coated layers, and also that between refractory carbide and carbon substrate. Therefore, the formation of a SiC concentration gradient on the surface of substrate carbon materials seems to be a very practical and

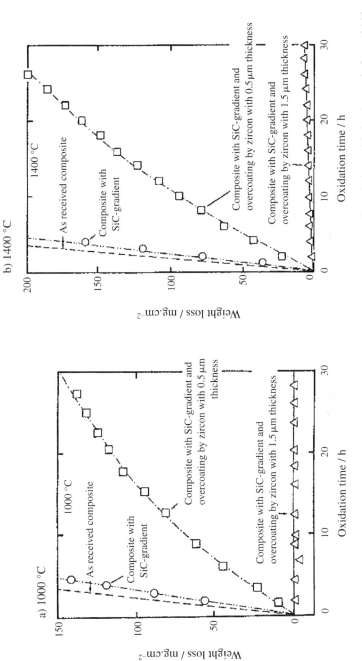

Figure 7-22 Weight loss of zircon-coated carbon/carbon composite with SiC concentration gradient during oxidation in an air flow of 200 ml min^{-1} [Yamamoto et al., 1995].

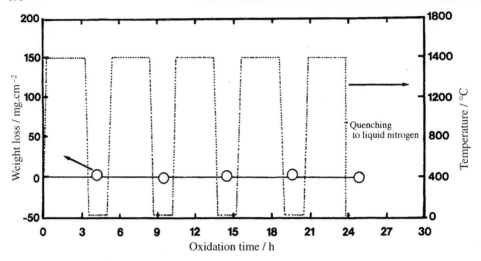

Figure 7-23   Weight loss of mullite-coated isotropic high-density graphite with SiC concentration gradient by rapid heating to 1400°C and quenching to room temperature [Yamamoto et al., 1993].

effective method to coat additional refractory oxide films, as shown in Figs. 7-22 and 7-23.

7.5. Carbon Fiber/Carbon Composites

(a) Architecture and performance

The two major components of carbon/carbon composites are the filler and the matrix carbons, the latter being the binder for the fillers, both of which can have different structures and textures. Based on variations and combinations of these two components, the properties of the carbon/carbon composites can be altered significantly. By the selection of processing conditions, additional variations are obtained. The representative filler for the composites is carbon fibers (carbon fiber/carbon composites, C/Cs), which give a wide range of variation in properties by selecting precursors and forms, either strand, yarn or chopped. The second major component of C/Cs is the matrix, the purpose of which is to assist in utilizing the high strength of the fibers, for example, by maintaining the space and geometries between yarns/fibers in order to transfer mechanical and heat loads between them. The structure and texture of matrix carbons are known to be the most important factors in determining the properties of C/Cs, such as fracture behavior, strength, toughness and thermal expansion. For example, the fracture behavior within the composite is controlled by the propagation of cracks within the matrix, at the fiber–matrix interface and across the fiber.

In Fig. 7-24, the relation between tensile strength and Young's modulus at room temperature is shown on various carbon materials, in order to specify the mechanical performance of C/Cs. The strength and modulus of C/Cs are almost comparable with carbon fibers of general-purpose grade (GP-grade), being far inferior to high-performance

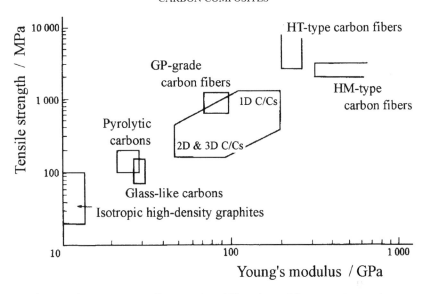

Figure 7-24   Relation between tensile strength and Young's modulus on various carbon materials.

grade (HP-grade) carbon fibers, but it should be emphasized that they have the highest values of strength and modulus among the bulk carbon materials. From the practical viewpoint of performance as structural materials, they are far superior to fibrous carbon fibers because C/Cs can be supplied in different sizes and shapes.

By controlling the architecture of the yarns and the structure/texture of the matrix, C/Cs can have the highest specific strengths and moduli of any known high-temperature materials, such as ceramics. This is possible because of the low specific density and high strength of carbon fibers, especially above 1500–2000°C. Although C/Cs are known primarily for their high temperature properties, they also can have unique thermal properties, such as zero coefficient of thermal expansion near room temperature and negative values below room temperature. Various problems on the interaction between filler fibers and matrix carbons have been discussed from the viewpoint of the changes in matrix structure and texture, in relation to stress graphitization [Inagaki and Meyer, 1999] using numerous examples, to illustrate how the mechanical and thermal properties of C/Cs could be extensively altered.

(b) Production

A flow sheet of the production of C/Cs is shown in Fig. 7-25. The forming before the densification process has been conducted mainly by impregnation, but also by filament winding in the case of continuous fibers passing through a precursor of matrix carbon (pitch or resin) and by injection molding of a mixture of matrix precursor with chopped carbon fibers. The densification process is essential to fill the pores formed during carbonization of these precursors in the matrix. It is the most important process to achieve high performance of C/Cs and is carried out either by chemical vapor deposition (CVD) of some hydrocarbon gases (or chemical vapor infiltration, CVI) or by impregnation of the

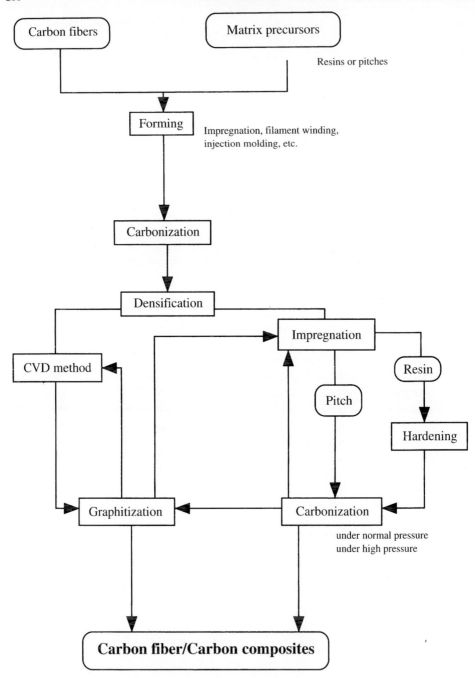

Figure 7-25   Flow sheet for the production of C/C composites.

organic precursor (pitch or resin). In the latter process, the impregnation of precursor and its carbonization has to be repeated, but in the former case carbonization and densification can be carried out in one unit process. The repetition of impregnation and carbonization of pitch was found to be effective in decreasing the content of open pores and, as a consequence, increasing the strength. It is clearly demonstrated as a relation between bending strength and open porosity in Fig. 7-26 [Tanabe et al., 1990].

It has also been pointed out that the insertion of the process of high-temperature heat treatment up to 2500°C is recommended to obtain high-density C/Cs. In Fig. 7-27, the bulk density of C/Cs is plotted against impregnation times as a function of heat treatment temperature [Furukawa et al., 1990]. Heat treatment at 2500°C is efficient in obtaining a high density. The selection of precursor for the carbon matrix is also an important factor for high performance. In Fig. 7-28, the bending strength of C/Cs is plotted as functions of the precursor and heat treatment temperature [Kimura et al., 1987]. Strength increases with increasing heat treatment temperature in general, but also strongly depends on the matrix.

The structural and textural development in the matrix has been discussed by emphasizing the mechanical interactions between the carbon fiber filler and the impregnated matrix [Hishiyama et al., 1974; Kimura et al., 1987; Inagaki and Meyer, 1999] because the impregnated precursors generally showed a large volume shrinkage

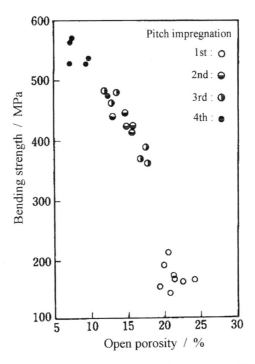

Figure 7-26   Changes in bending strength and content of open pores with repetition of the cycles of pitch impregnation and carbonization to prepared C/C with unidirectionally aligned carbon fibers [Tanabe et al., 1990].

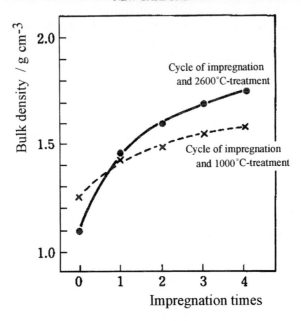

Figure 7-27   Changes in bulk density of C/Cs with impregnation time of pitch combined with intermittent heat treatment at two different temperatures, 1000 and 2600°C [Furukawa et al., 1990].

during carbonization, although the filler carbon fibers did not because they were already carbonized. The boundaries between filler carbon fibers and the matrix formed from a thermosetting resin, which gives glass-like carbon without fillers, are shown in Fig. 7-29. The structural change to graphite and the orientation of layer planes along the periphery of carbon fibers are clearly demonstrated. Graphitization of the matrix carbon was also shown by X-ray diffraction measurement. This structural change at the boundary in C/Cs was supposed to be due to the relaxation of accumulated stress during carbonization and graphitization in these C/Cs due to a large shrinkage of the matrix and was discussed in relation to stress graphitization [Inagaki and Meyer, 1999].

(c) Mechanical behavior

Since the application of C/Cs is considered mostly as the structural components, particularly at high temperatures, mechanical and thermal properties have been studied intensively. The mechanical strength of most carbon materials has been known to increase with increasing temperature up to around 2500°C [Malmstrom et al., 1951; Martens et al., 1958; Inagaki et al., 1963], as can be seen in Fig. 7-19. They are known to show plastic deformation at high temperatures, above 2000°C, but rather brittle fracture at low temperatures. By making composites with carbon fibers, therefore, one expects to modify their fracture mode from brittle to ductile even at low temperatures; in other words, to produce a toughness, in addition to increasing their breaking strength.

It is known that the fracture mode is quite different between low-temperature-treated carbon, which has a glass-like carbon texture, and graphitized carbon, which consists of anisotropic hexagonal carbon layers; in the former a single crack usually propagates through the specimen and leads to instant macroscopic breaking, but in the latter multiple cracks with varying lengths occur and rather ductile fracture behavior is observed. In C/Cs, therefore, the matrix texture has a great influence on the fracture behavior. In composites with a glass-like carbon matrix, a single type of fracture that usually occurs at the fiber–matrix boundary travels unimpeded through the fibers and surrounding matrix. In a graphitic matrix, in contrast, multiple fractures occur in the matrix, depending primarily on the texture and orientation of the lamellar structure of the matrix, and failure usually takes place when these cracks join together to form a critical size. The difference in fracture mode in these two kinds of matrix makes a great difference to the mechanical properties of the C/Cs. In addition, the strain characteristics of C/Cs are very different, being low for glass-like carbon and relatively high for the graphite type of matrix.

The effect of texture on mechanical and thermal properties has been demonstrated by various C/Cs. In Fig. 7-30a, the relative toughness of a C/C, composed of three-dimensional woven PAN-based carbon fibers and densified with a pitch-based matrix, is plotted against HTT [Meyer, 1986]. The matrix texture was observed to be ungraphitized

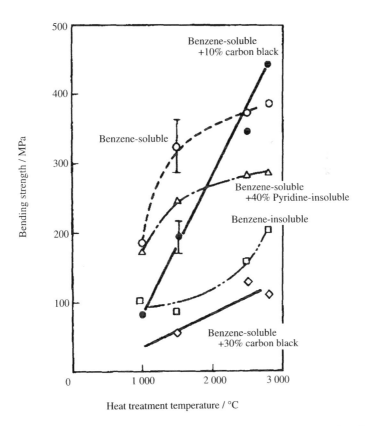

Figure 7-28 Dependence of bending strength on heat treatment temperature and on the precursor of matrix carbon [Kimura et al., 1987].

Figure 7-29 SEM micrograph of the boundary between filler carbon fibers and matrix carbon which was derived from a thermosetting resin. [Courtesy of Prof. Yasuda of Tokyo Institute of Technology.]

carbon in the C/C heat-treated at 2100°C, a mixture of carbon and graphite at 2400°C and finally a highly graphitic matrix above 2500°C. The toughness value is a maximum at HTT of 2400°C, and then decreased at 2550 and 2750°C. This result is explained by the change of crack formation in the matrix of C/Cs, as illustrated in Fig. 7-30b. The mode of crack propagation in the matrix can vary for differences in texture. In ungraphitized carbon (b-1 in Fig. 7-30), a single type of crack propagates along the matrix with high modulus and isotropic texture. Since not much energy is dissipated during the propagation of this single type crack, the toughness of the C/C is low. The matrix, which has been heat-treated at 2400°C and composed of two regions of ungraphitized and graphitized carbon, has anisotropic modulus value and facilitates the formation of multiple small cracks that absorb more energy (b-2); as a consequence, the toughness of this type of matrix texture becomes higher value than the previous one (b-1). As the matrix texture becomes highly graphitic with longer lamellae and very anisotropic modulus values after heat treatment above 2500°C, however, many large cracks can easily propagate along the matrix and less energy is expended (b-3). Therefore, the toughness of the composite with this highly graphitic texture is lower than the previous one (b-2). The change in modulus of the matrix in C/Cs with HTT has been shown experimentally [Feldman and Gyetvay, 1986].

The selection of the correct conditions of heat treatment is very important for carbon fiber/glass-like carbon matrix composites to obtain the desired mechanical properties. Figure 7-31 shows the relations between load and displacement during a four-point bending test on these composites with different heat treatment temperatures [Kimura and Yasuda, 1983]. The carbon fibers were aligned along the longitudinal direction of the sample. After the initial elastic deformation segment, the composites that have been heat-treated above 2500°C have a plastic displacement, as shown by the portions of A–B and C–D of the curves, which occurs before the formation of macroscopic cracking in the specimen. In contrast, the composite heat-treated at a low temperature such as 1000°C does not have such a plastic-like portion. By combination with microscopic observation of

(a)

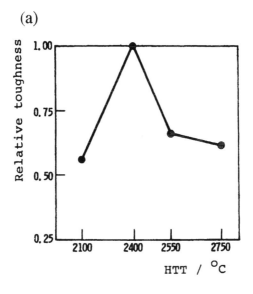

(b)

(b-1)

(b-2)

(b-3)

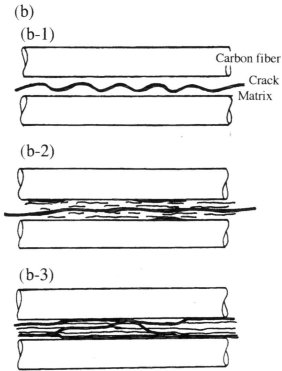

Figure 7-30 Dependence of relative toughness on HTT (a) and schematic illustration of crack propagation in the matrix for the C/Cs of three-dimensional woven PAN-based carbon fibers densified with pitch impregnation. [Courtesy of Prof. Meyer.]

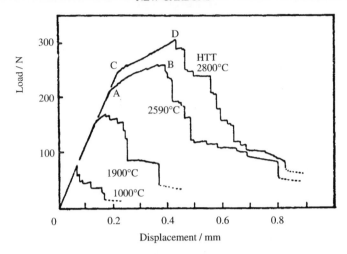

Figure 7-31 Load–displacement curves on C/Cs as a function of HTT. [Courtesy of Prof. Yasuda of Tokyo Institute of Technology.]

textures of the matrix in the composites, this behavior was attributed to the formation of a lamellar texture in the matrix, as shown in Fig. 7-29, where microscopic cracks are formed in this graphitized region near fibers, thereby causing stress relaxation and a redistribution of the local mechanical forces to occur. This microcracking decreases the sample's modulus and increases its plastic deformation. Ultimately, the microcracks join together to form critical size macrocracks and cause failure, as occurs at points B and D of the curves, with jagged steps of declining strength.

Figure 7-32 shows that the texture changes in the matrix with heat treatment have a sensitive effect on mechanical behaviors by using 2D-C/Cs, composed of PAN-based carbon fibers and phenolic resin, and heated to 2480°C [Buechler et al., 1984]. In the as-received sample A with a simple heat treatment at 2480°C, failure occurs in a brittle manner at a deflection of about 1.44 mm. In the same type of C/C (sample B), which has had additional heat treatment at about 2480°C, however, the failure point increases by 45% to a deflection value of about 1.99 mm, and clearly a more ductile type of failure has taken place. In comparison, another sample which has undergone a much longer period of heat treatment (sample C) shows a gradual deflection to 2.54 mm, which is clearly a plastic mode of failure. SEM examination of the matrix surrounding the fibers on these three samples showed that the composite with the longer period of heat treatment contained a greater amount of the graphitic component.

In order to show the importance of matrix texture, in Fig. 7-33 the relations between load and loadpoint displacement are compared with two C/Cs of which a matrix has been formed by different methods, and also with polycrystalline graphite (isotropic high-density graphite) [Sakai et al., 1991; Miyazima and Sakai, 1992]. The sample 2D-C/C was prepared by the carbonization of a carbon fiber/phenol resin composite with two-directional alignment of fibers followed by the repetition of pitch impregnation, and the sample felt-C/C was prepared by the densification of carbon fiber felt with chemical vapor infiltration. The former shows much higher strength and larger strain than other two. On

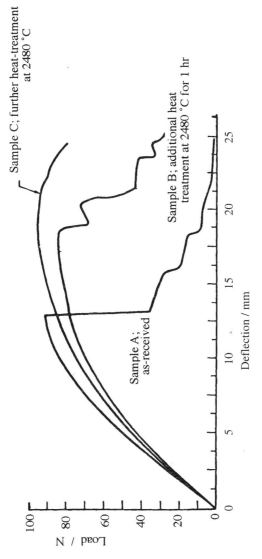

Figure 7-32 Load–deflection curves on C/Cs with different heat treatment times. [Courtesy of Prof. Meyer.]

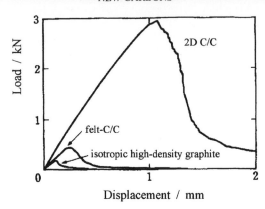

Figure 7-33 Load–loadpoint displacement relations on three different carbon materials [Sakai et al., 1991].

the fractured surface of sample 2D-C/C, marked pull-out of carbon fibers is seen, as shown in Fig. 7-34. However, no carbon fiber pull-out was observed on sample felt-C/C, and therefore the increase in strength was not pronounced. The difference among these three samples in mechanical behavior could be shown by measuring energy to break (work of fracture); the present 2D-C/C had a large value of about $6000 \, J \, m^{-2}$, but felt-C/C only $275 \, J \, m^{-2}$ and polycrystalline graphite a much lower value of $80 \, J \, m^{-2}$ [Sakai et al., 1991].

From these detailed measurements of fracture behavior, processes leading to fracture in C/Cs were discussed [Sakai, 1995]. Figure 7-35 illustrates these processes schematically, showing a single filament of carbon fiber. With increasing load, the matrix breaks first because it has much lower strength than carbon fiber, and the crack reaches the fiber, but at this load the fiber does not break because of its high strength. Further increase in load causes breaking at the interface between fiber and matrix, because this part is usually much weaker than the fiber itself, and the crack then propagates beyond the fiber. Finally, the carbon fiber itself breaks somewhere and then macroscopic fracture of the composite

Figure 7-34 SEM micrographs on the fracture surface of C/Cs, showing fiber pull-out (a) and no pull-out (b). [Courtesy of Prof. Sakai of Toyohashi University of Technology.]

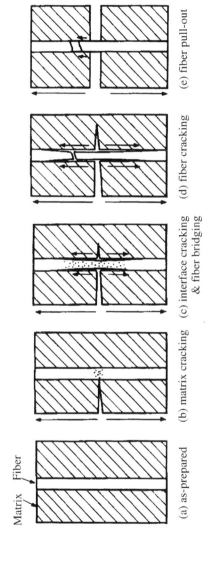

Figure 7-35 Processes for the fracture of C/Cs. [Courtesy of Prof. Sakai of Toyohashi University of Technology.]

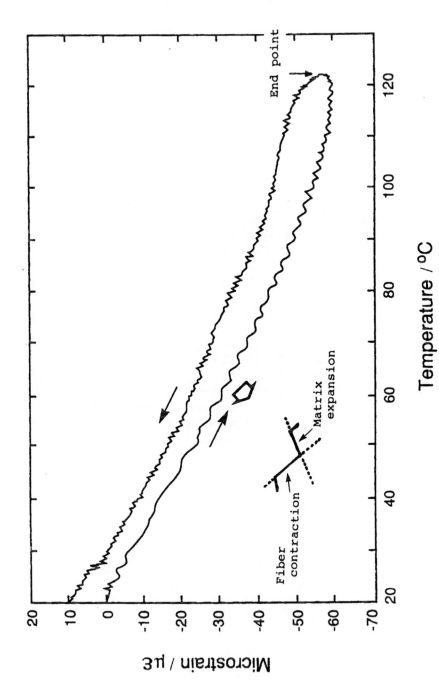

Figure 7-36 Microdeformation in the length of a thin-walled C/C tube during a cycle of heating to 121°C and cooling to room temperature. [Courtesy of Prof. Meyer.]

occurs. Since the breaking of the carbon fiber does not necessarily occur at the same place as where the crack in the matrix arrives, it is well understood that fiber pull-out is observed. In this scheme, two processes, fiber bridging due to interface cracking and fiber pull-out, are characteristic for C/Cs and are found to spend a large part of energy for fracture (work of fracture). This suggests the importance of controlling the structure of the interface between the matrix carbon and filler carbon fiber, in addition to controlling the structure of the matrix carbon and selection of filler carbon fibers.

(d) Thermal expansion

The bond interface between the fiber and its surrounding matrix is an important factor because it determines the thermal properties of C/Cs, as well as the mechanical properties. Details of the bond interface were studied by high-resolution transmission microscopy [Pleger et al., 1991].

Along the fiber axis, a shear force is assumed to develop with increasing temperature, because the thermal expansion coefficient of the fibers is negative but that of the matrix is positive. In certain cases, therefore, this shear stress at the filler–matrix interface can be large enough to break the bonds and then the matrix dominates the thermal expansion of C/Cs [Meyer, 1989]. However, only a fraction of the bonds is usually broken in the C/Cs during thermal cycling. Accordingly, the thermal stress loads are redistributed or relaxed so that the expansion/contraction of the sample occurs incrementally during thermal cycling. In Fig. 7-36, the length of thin-walled C/C tubes is continuously measured, with a precision of 0.2 $\mu\varepsilon$ units, where ε is the change in the length of the tube relative to the original, during the cycle between 121°C and room temperature.

The saw-tooth pattern is considered to be indicative of the breaking bonds at filler–matrix interfaces. As the temperature increases, the negative slope of the pattern is due to the axial contraction of the fiber, whereas the positive slope is caused by the expansion of the matrix, as illustrated schematically in the figure. The non-uniform nature of the saw-tooth pattern is attributed to the variability in the shear stresses that are created or destroyed during the sample's heating and cooling cycle. The amplitude of this pattern is between 1.4 and 2.1 $\mu\varepsilon$, or 7–10 times that of the precision of these measurements [Meyer and Henderson, 1994].

The thermal behavior of C/Cs is dependent on the interaction between adjacent fibers through the influences of the textural characteristics of the surrounding matrix and the filler–matrix interfaces. It was also demonstrated that the texture of the matrix and the state of filler–matrix bonding determined the thermal expansion, and that there can be a combination of ungraphitized and oriented graphitized regions that enable the optimum thermal properties for specific engineering applications to be obtained [Inagaki and Meyer, 1999].

CHAPTER 8

Concluding Remarks

In the present book, carbon materials which have been developed recently and/or for which new applications have been found, i.e. *new carbons*, were reviewed in six chapters: highly oriented graphite, isotropic carbons of glass-like carbons and high-density graphites, carbon fibers, porous carbons, intercalation compounds and carbon composites. In Table 8-1, some characteristics of the structure and texture of these new carbons described in this book are summarized with their functions, together with other carbon materials, classical carbons of graphite electrodes, carbon blacks and activated carbons, all of which consist principally of an sp^2 hybrid orbital between carbon atoms, and also carbon allotropes based on different carbon–carbon bond natures of sp, sp^2 and sp^3 hybrid orbitals.

The carbon materials described in the present book are based on sp^2 hybrid orbital and graphitic structures; in other words, the fundamental structural units are hexagonal carbon layers. In principle, therefore, they are anisotropic. This anisotropy sometimes causes some difficulty in understanding and using them; however, it provides a certain variety in these carbon materials. As explained in Chapter 1 (Fig. 1-10), the preferred orientation of these anisotropic structural units makes various morphologies of carbon materials possible; flat plates due to planar orientation, fibrous carbons due to axial orientation with either radial or co-axial arrangement, small round particles with either radial or concentric point orientation of structural units, and also particles with an isotropic nature due to random orientation of minute structural units. There are various intermediate textures among these orientation schemes; for example, PAN-based carbon fibers have rather well-oriented structural units along their fiber axis, but in their cross-sections structural units are arranged rather randomly in their HT type and have a little better organization in their HM type, but far from well-defined co-axial arrangement in vapor-grown carbon fibers, as explained in Chapter 4. In the particles of different cokes, a similar situation is observed; in so-called needle-like coke particles structural units are oriented in a planar scheme, but are more random in so-called regular cokes, although both of them are intermediate between two extreme cases, perfect planar and random orientations. This variety of nanotexture in particles leads to another variety of textures in micrometric and millimetric scales; for example, the mechanical performance of carbon/carbon composites is predominantly governed by the combination of carbon fibers and matrix carbons: the oriented nanotexture of matrix carbon gives toughness, as explained in Chapter 7 (Section

7.5). This nanotexture has a strong effect on the intercalation reactions of various alkali metals and metal halides, and governs the properties of the resultant intercalation compounds, as shown in Chapter 6.

Table 8-1. Characteristics of structure, texture and functions of various carbon materials

Carbon materials	Structure and texture	Functions	Chapter
Classical Carbons			
Graphite electrodes	Large size	Electrically conductive Thermal shock resistant	
Carbon blacks	Minute primary particles	High compatibility with organic materials	
	Different modes of particle aggregation	Black color	
Activated carbons	Porous	Adsorption ability Supporting ability	
New Carbons			
Highly oriented graphites	High graphitization degree Pronounced planar orientation	High anisotropy	2
Isotropic carbons			
Glass-like carbons	Random orientation of structural units	Gas impermeability Hard Low bulk density	3
High-density graphite	Random orientation High density	High purity Easy machining	3
Carbon fibers	Axial orientation of structural units Various cross-sectional textures	Fibrous morphology High mechanical performance	4
Porous carbons	Porous High surface area	Adsorption ability Supporting ability	5
Intercalation compounds	Nanoscale composites Variety in the combination between host and guest	High electrical conductivity Electrochemical activity Functional nanospace	6
Carbon composites	Micro- and macroscale composites Variety in combinations	Variety in applications	7
Carbon Allotropes			
Fullerenes	Spherical shape Closest packing	Molecular nature Two modes of doping	
Diamond	Covalent C–C bond by sp^3 hybrid orbital From crystalline to amorphous	High hardness Transparency Electrical insulator	
Carbyne	sp hybrid orbitals Chain structure		

developments in carbon materials with such a wide range of structures, textures and properties, and also strong demands from modern technology related to materials science, new concepts and strategies in carbon research need to be developed. A new strategy, *carbon alloys*, was employed by a Japanese carbon group, which has been working since 1997 on major research projects for the Ministry of Education, Grant-in-Aid for Scientific Research on Priority Areas, "Carbon Alloys" [Tanabe and Yasuda, 2000].

Carbon alloys were tentatively defined as follows [Yasuda, 1995; Tanabe and Yasuda, 2000]: "Carbon alloys are materials mainly composed of carbon atoms in multicomponent systems, in which each component has physical and/or chemical interaction with each other. Here, carbons with different hybrid orbitals account as different components".

Boron has been known to substitute for the carbon atom in carbon materials in each family, which can be called alloying according to the exact definition of alloys and also called a solid solution according to the definition of crystal chemistry. If one applies the definition described above, however, most carbon materials are carbon alloys, as can be understood easily. In addition, the concept of carbon alloys provides new possibilities for combination with other materials, e.g. B-C-N compounds with either sp^2, sp^3 or their mixture, new understanding of the combination between metal carbide and carbon, etc. In this definition of carbon alloys, porous carbons can be understood as the combination between carbon and pores. In the Japanese research project, special attention is paid to rear (latent) surfaces, which may give a bimodal function to carbon materials.

This concept led to attention being paid to nano- and micro-sized spaces, which are expected to be formed in different carbon alloys. Another research project on "Creation of functional nano- and micro-sized spaces in carbon materials" has been conducted since 1996 by the Research for the Future Program of Japan Society for the Promotion of Science (JSPS). A few research projects are also being undertaken: water purification using carbon fibers, recovery and recycling of heavy oil using carbon materials and SOx elimination from exhaust gases using activated carbon fibers, by the New Energy and Industrial Technology Development Organization (NEDO).

The new and interesting results of these research projects by Japanese groups will be published in different journals. They are expected to promote our understanding of a wide variety of carbon materials, to develop new functions and applications of these carbon materials, and to make certain contributions to various global problems, including the environment, energy and resources.

Acknowledgements

The author would like to express his sincere thanks to Prof. Y. Hishiyama of Musashi Institute of Technology for his kind cooperation in writing this book and also Mme A. Oberlin of Directeur de Recherche CNRS Emeritus for her kind advice on this book. He would like to thank those people who kindly provided their pictures for this book, and also a sincere thanks to Charlotte Harding for the copy-editing.

References

Abe, Y., Imamura, R., Yoshida, S. and Oya, A., *Tanso*, **1996** [No.172], 11, 1996.
Akihama, S., Suenaga, T. and Banno, T., *KICT Report* No.53, 1984.
Akihama, S., *Sen-i Gakkai Shi*, **44**, 125, 1988.
Audier, M., Oberlin, A., Oberlin, M., Coulon, M. and Bonnetain, L., *Carbon*, **19**, 217, 1981.
Auguie, D., Oberlin, M., Oberlin, A. and Hyvernay, P., *Carbon*, **18**, 337, 1980.
Avnir, D., Farin, D. and Pfeifer, P., *J. Chem. Phys.*, **79**, 3566, 1983.
Bacon, B.E., *Acta Cryst.*, **4**, 392, 1951.
Bahl, O.P., "*Carbon Fibers*", 3rd Edition, J.-B. Donnet et al., (eds), Marcel Dekker, New York, p.1, 1998.
Baker, R.T.K., Terry, S. and Harris, P.S., *Nature*, **253**, 37, 1975.
Baker, D.F. and Bragg, R.H., *J. Non-Cryst. Solids*, **53**, 57, 1983.
Basu, S., Zeller, C., Flander, P.J., Fuerst, C.D. Johnoson, W.D. and Fisher, J.E., *J. Mater. Sci. Eng.*, **38**, 275, 1979.
Beck, F. and Krohn, H., *Synth. Met.*, **7**, 193, 1983.
Berger, D. and Maire, J., *Mater. Sci. Eng.*, **31**, 335, 1977.
Blackman, L.C.F. and Ubbelohde, A.R., *Proc. Roy. Soc., London*, **266**, 20, 1962.
Bokros, J., "*Chemistry and Physics of Carbon*", Vol. **5**, P.L. Walker, (ed.), Marcel Dekker, New York, p.3, 1969.
Bokros, J., Lagrange, L.D. and Shoen, F.J., "*Chemistry and Physics of Carbon*", Vol. **9**, P.L. Walker, (ed.), Marcel Dekker, New York, p.103, 1972.
Bourgerette, C., Oberlin, A. and Inagaki, M., *J. Mater. Res.*, **7**, 1158, 1992.
Bourgerette, C., Oberlin, A. and Inagaki, M., *J. Mater. Res.*, **8**, 121, 1993.
Bourgerette, C., Oberlin, A. and Inagaki, M., *J. Mater. Res.*, **10**, 1024, 1995.
Bowman, J.C., *Proc. 1st and 2nd Conf. on Carbon*, Buffalo Univ., New York, p.59, 1956.
Brooks, J.D. and Taylor, G.H., "*Chemistry and Physics of Carbon*", Vol. **4**, P.L. Walker, (ed.), Marcel Dekker, New York, p.503, 1968.
Buechler, M., Hawkins, G.F. and Meyer, R.A., *Aerospace Rpt. TOR-0084 (4622-01)*, 1984.
Bunsell, A.R., "*Fibre Reinforcements for Composites Materials*", Elsevier, Amsterdam, 1988.
Cao, N.Z., Shen, W.C., Wen, S.Z., Gu, J.L. and Wang, Z.D., *Extended Abstracts of Carbon '96*, Newcastle upon Tyne, p.112, 1996.
Chambers, A., Park, C., Baker, R.T.K. and Rodriguez, N.M., *J. Phys. Chem. B*, **102**, 4253, 1998.
Dahn, J.R., Sleigh, A.K., Shi, H., Way, B.M., Weydanz, W.J., Reimers, J.N., Zhong, Q. and von Sacken, U., "*Lithium Batteries*", G. Pistoia, (ed.), p.1, 1994.
Dahn, J.R., *22nd Biennial Conference on Carbon*, San Diego, 1995.
Daulan, C., Derre, A., Flandrois, S., Roux, J.C. and Saadaoui, H., *J. Phys. I France*, **5**, 1111, 1995.
Daumas, N. and Herold, A., *Compt. Rend.*, **268**, C-373, 1969.
de Fonton, S., Oberlin, A. and Inagaki, M., *J. Mater. Sci.*, **15**, 909, 1980.
Derbyshire, F., Jagtoyen, M. and Thwaites, M., "*Porosity in Carbons*", J.W. Patrick, (ed.), Edward Arnold, London, p.227, 1995.
Despres, J.F., Genseki, A. and Odawara, O., *J. Mater. Chem.*, **7**, 1877, 1997.
Donnet, J.-B., Papier, E., Wang, W. and Stockli, F., *Carbon*, **32**, 183, 1994.
Donnet, J.-B., Wang, T.K., Rebouillat, S. and Peng, J.C.M., "*Carbon Fibers*", 3rd Edition, Marcel Dekker, New York, 1998.
Dresselhaus, M.S. and Dresselhaus, G., *Adv. Phys.*, **30**, 139, 1981.
Edie, D.D., Fain, C.C., Robinson, K.E., Harper, A.M. and Rogers, D.K., *Carbon*, **31**, 941, 1993.
Edie, D.D., Robinson, K.E., Fleurot, O., Jones, S.P. and Fain, C.C., *Carbon*, **32**, 1045, 1994.
Egashira, M., Katsuki, H., Hayashi, K. and Kawazumi, S., *Sekiyu Gakkai Shi*, **26**, 247, 1983.

Ehrburger, P., Baranne, P. and Lahaye, J., *Carbon*, **24**, 495, 1986.

Endo, M., Koyama, T. and Hishiyama, Y., *J. Appl. Phys.*, **15**, 2073, 1976.

Endo, M., *Ph.D. Thesis*, Nagoya Univ., 1978.

Endo, M., Mori, T., Koyama, T. and Inagaki, M., *Ouyou Butsuri*, **49**, 563, 1980.

Endo, M. and Shikata, M., *Ouyou Butsuri*, **54**, 507, 1985.

Endo, M., Inagaki, M. and Dresselhaus, M.S., *"Advanced Carbon Series II, Graphite Intercalation Compounds"*, Carbon Society of Japan, p.129, 1990.

Endo, M., Nakamura, H., Emori, A., Ishida, T. and Inagaki, M., *Tanso*, **1991** [No.150], 319, 1991.

Endo, M., Takeuchi, K., Kobori, K., Takahashi, K., Kroto, H.W. and Sarkar, A., *Carbon*, **33**, 873, 1995.

Endo, M., Nishimura, Y. Takahashi, T. and Dresselhaus, M.S., *J. Phys. Chem. Solids*, **57**, 725, 1996.

Endo, M., Kim, C., Nishimura, K., Fujino, T. and Miyashita, K., *Carbon*, **38**, 183, 2000.

Eto, M., Arai, T. and Konishi, T., *JAERI-Res.*, **98-024**, 1998.

Feldman, L.A. and Gyetvay, S.R., *Aerospace Rpt. TOR-0086(6728-020)-1*, 1986.

Fitzer, E., Schaffer, W. and Yamada, S., *Carbon*, **7**, 643, 1969.

Flandrois, S., Masson, J.M., Rouillon, C.J., Gaultiere, J. and Hauw, C., *Synth. Met.*, **3**, 1, 1981.

Franklin, R.E., *Acta Cryst.*, **4**, 253, 1951.

Fujii, R. and Dohi, T., *"Graphite Intercalation Compounds"*, Chapter 5, N. Watanabe, ed., Kindai-Heshu-Sha, Tokyo, p.281, 1986.

Fujiie, K., Minagawa, S., Suzuki, T. and Kaneko, K., *Chem. Phys. Lett.*, **236**, 427, 1995.

Fujimaki, H. and Otani, S., *Ceramics*, **11**, 612, 1976.

Furdin, G., Langrange, P. and Herold, A., *C.R. Acad. Sci. Paris*, **283**, C-563, 1976.

Furukawa, S., Otani, S., Kojima, A. and Kodaira, I., *Tanso*, **1990** [No.141], 23, 1990.

Gaier, J.R. and Jaworske, D.A., *Synth. Met.*, **12**, 525, 1985.

Gilorami, L., *Fuel*, **42**, 229, 1963.

Gole, J., *J. Mater. Sci. Eng.*, **31**, 309, 1977.

Green, L., *J. Appl. Phys.*, **20**, 289, 1953.

Guerard, D. and Herold, A., *Carbon*, **13**, 337, 1975.

Guigon, M., Oberlin, A. and Desarmot, G., *Fibre Sci. Technol.*, **20**, 55, 1984a.

Guigon, M., Oberlin, A. and Desarmot, G., *Fibre Sci. Technol.*, **20**, 177, 1984b.

Hatori, H., Yamada, Y. and Shiraishi, M., *Carbon*, **30**, 303, 1992a.

Hatori, H., Yamada, Y. and Shiraishi, M., *Carbon*, **30**, 305, 1992b.

Hatori, H., Yamada, Y. and Shiraishi, M., *Carbon*, **30**, 763, 1992c.

Hatori, H., Yamada, Y,, Shiraishi, M., Nakada, H., Yoshitomi, S., Yoshihara, M. and Kimata, T., *Tanso*, **1995** [No.167], 94, 1995a.

Hatori, H., Yamada, Y. and Shiraishi, M., *J.Appl. Polym. Sci.*, **57**, 871, 1995b.

Hishiyama, Y., Inagaki, M., Kimura, S. and Yamada, S., *Carbon*, **12**, 249, 1974.

Hishiyama, Y. and Kaburagi, Y., *Tanso*, **1979** [No.98], 89, 1979.

Hishiyama, Y., Yoshida, A. and Inagaki, M., *Carbon*, **20**, 79, 1982.

Hishiyama, Y., Kaburagi, Y. and Yoshida, A., *"Sciences and New Applications of Carbon Fibers"*, Toyohashi Univ., p.21, 1984.

Hishiyama, Y. and Ono, A., *Carbon*, **23**, 445, 1985.

Hishiyama, Y., *Tanso*, **1987** [No.130], 124, 1987.

Hishiyama, Y. and Kaburagi, Y., *Tanso*, **1987** [No.128], 18, 1987.

Hishiyama, Y., Kaburagi, Y. and Inagaki, M., *"Chemistry and Physics of Carbon"*, Vol. 23, P.A. Thrower, (ed.), Marcel Dekker, New York, p.1, 1991.

Hishiyama, Y., *Hyoumen*, **30**, 473, 1992.

Hishiyama, Y., Yoshida, A., Kaburagi, Y. and Inagaki, M., *Carbon*, **30**, 333 1992.

Hishiyama, Y., Kaburagi, Y. and Yoshida, A., *Carbon*, **31**, 777, 1993a.

Hishiyama, Y., Yoshida, A. and Kaburagi, Y., *Carbon*, **31**, 1265, 1993b.

Hishiyama, Y., Nakamura, M., Nagata, Y. and Inagaki, M., *Carbon*, **32**, 645, 1994.

Hishiyama, Y., Igarashi, K., Kanaoka, I., Fujii, H., Kaneda, T., Koidesawa, T., Shimazawa, Y. and Yoshida, A., *Carbon*, **35**, 657, 1997.

Hishiyama, Y., Yoshida, A. and Inagaki, M., *Carbon*, **36**, 1113, 1998.

Hishiyama, Y., Private Communication, 1998.

Honda, H., Kimura, H. and Sanada, Y., *Carbon*, **9**, 695, 1971.

Honda, H., Yamada, Y., Oi, S. and Fukuda, K., *Tanso*, **1973** [No. 72], 3, 1973.

Houska, C.R. and Warren, B.E., *J.Appl. Phys.*, **25**, 1503, 1954.

Huttepain, M. and Oberlin, A., *Carbon*, **28**, 103, 1990.

Iijima, S., *Nature*, **354**, 56, 1991.

Inagaki, M., *Tanso*, **1963** [No.34], 18, 1963a.

Inagaki, M., *Tanso*, **1963** [No.36], 16, 1963b.
Inagaki, M., Takeuchi, H., Yamanaka, F. and Noda, T., *Kogyou Kagaku Kaishi*, **66**, 169, 1963.
Inagaki, M., *Tanso*, **1968** [No.53], 61, 1968.
Inagaki, M., Furuhashi, H., Ozeki, T. and Naka, S., *J. Mater. Sci.*, **8**, 312, 1973.
Inagaki, M., *Tanso*, **1978** [No.94], 106, 1978.
Inagaki, M., Kuroda, K. and Sakai, M., *High Temp.-High Press.*, **13**, 207, 1981.
Inagaki, M., Muramatsu, K. and Maeda, Y., *Synth. Met.*, **8**, 335, 1983a.
Inagaki, M., Matsumoto, A., Sakai, M. and Maeda, Y., *Nihon Kagaku Kaishi*, **1983**, 309, 1983b.
Inagaki, M., "*Funryutai Kougaku*", Maki-Shoten, p.75, 1985a.
Inagaki, M., "*Tanso Zairyo Kougaku*", Nihon Kougyou Shinbunsya, 1985b.
Inagaki, M., *Tanso*, **1985** [No.122], 114, 1985c.
Inagaki, M., "*Chemical Physics of Intercalation*", A.P. Legrand and S. Flandrois, (eds), p.105, 1987.
Inagaki, M., *Tanso*, **1988** [No.133], 127, 1988.
Inagaki, M., *J. Mater. Res.*, **4**, 1560, 1989.
Inagaki, M., Harada, S., Sato, T., Nakajima, T., Horino, Y. and Morita, K., *Carbon*, **27**, 253, 1989.
Inagaki, M., Itoh, E. and Tanaka, A., *Synth. Met.*, **35**, 383, 1990a.
Inagaki, M., Iwashita, N. and Kouno, E., *Carbon*, **28**, 49, 1990b.
Inagaki, M., *Carbon*, **29**, 287, 1991.
Inagaki, M., Iwashita, N., Hishiyama, Y., Kaburagi, Y., Yoshida, A., Lafdi, K., Bonnamy, S. and Oberlin, A., *Tanso*, **1991**, 57, 1991a.
Inagaki, M., Sakakmoto, K. and Hishiyama, Y., *J. Mater. Res.*, **6**, 1108, 1991b.
Inagaki, M. and Wang, Z.D., *Tanso*, **1992**, 184, 1992.
Inagaki, M., Sato, M., Takeichi, T., Yoshida, A. and Hishiyama, Y., *Carbon*, **30**, 903, 1992a.
Inagaki, M., Ishida, T., Yabe, K. and Hishiyama, Y., *Tanso*, **1992**, 244, 1992b.
Inagaki, M., Ibuki, T. and Takeichi, T., *J. Appl. Polym. Sci.*, **44**, 521, 1992c.
Inagaki, M. and Ohira, M., *Carbon*, **31**, 777, 1993.
Inagaki, M. and Iwaswhita, N., *J. Power Sources*, **52**, 69, 1994.
Inagaki, M. and Hishiyama, Y., "*New Carbon Materials*", Gihoudou Shuppan, Tokyo, 1994.
Inagaki, M., Hishiyama, Y. and Kaburagi, Y., *Carbon*, **32**, 637, 1994.
Inagaki, M., *Tanso*, **1995** [No.170], 298, 1995.
Inagaki, M. and Sunahara, M., *Tanso*, **1998** [No.183], 146, 1998.
Inagaki, M., Tachikawa, H., Nakahashi, T. Konno, H. and Hishiyama, Y., *Carbon*, **36**, 1021, 1998.
Inagaki, M., "*World of Carbon*" Vol. **1**, P. Delhaes, (ed.) (in press), 2000.
Inagaki, M. and Hishiyama, Y., "*World of Carbon*" Vol. **1**, P. Delhaes, (ed.) (in press), 2000.
Inagaki, M. and Meyer, R.A., "*Chemistry and Physics of Carbon*", Vol. **26**, P.A. Thrower and L. Radovic, (eds), p.149, 1999.
Inagaki, M., Hishiyama, Y., Takeichi, T. and Oberlin, A., "*Chemistry and Physics of Carbon*", Vol. **26**, P.A. Thrower and L. Radovic, (eds), p.246, 1999a.
Inagaki, M., Vignal, V., Konno, H. and Morawski, A.W., *J. Mater. Res.*, **14**, 3152, 1999b.
Ishioka, M., Okada, T. and Matsubara, K., *Carbon*, **30**, 859, 1992a.
Ishioka, M., Okada, T., Matsubara, K. and Endo, M., *Carbon*, **30**, 865 1992b.
Ishioka, M., Okada, T. and Matsubara, K., *Carbon*, **30**, 975, 1992c.
Ishioka, M., Okada, T. and Matsubara, K., *Carbon*, **31**, 123, 1993a.
Ishioka, M., Okada, T. and Matsubara, K., *Carbon*, **31**, 699, 1993b.
Ishioka, M., Okada, T., Matsubara, K., Inagaki, M. and Hishiyama, Y., *J. Mater. Res.*, **8**, 1866, 1993c.
Ishioka, M., Hishiyama, Y. and Inagaki, M., *Tanso*, **1995**, 218, 1995.
Isoda, S., Shimada, H., Kochi, M. and Kambe, H., *J. Polym. Sci., Polym. Phys.*, **19**, 1293, 1981.
Iwashita, N. and Inagaki, M., *Synth. Met.*, **34**, 139, 1989.
Iwashita, N., *Ph.D. Thesis*, Hokkaido University, 1992.
Iwashita, N. and Inagaki, M., *Nihon Kagaku Kai Shi*, **1992**, 1414, 1992.
Iwashita, N. and Inagaki, M., *Carbon*, **31**, 1107, 1993.
Iwashita, N., Inagaki, M. and Sawada, Y., *Osaka Res. Inst.*, **44**, 115, 1993.
Iwashita, N. and Inagaki, M., *Solid State Ionics*, **70/71**, 425, 1994.
Iwashita, N., Inagaki, M. and Hishiyama, Y., *Carbon*, **35**, 1073, 1997.
Jenkins, G.M. and Kawamura, K., *Nature*, **231**, 175, 1971.
Joy, D.C., Newbury, D.E. and Davidson, D.L., *J. Appl. Phys.*, **53**, R81, 1982.
Kaburagi, Y., Hishiyama, Y., Baker, D.F. and Bragg, R.H., *Phil. Mag. B*, **54**, 381, 1986.
Kaburagi, Y., Yoshida, A., Hishiyama, Y., Nagata, Y. and Inagaki, M., *Tanso*, **1995**, 19, 1995.
Kaneko, K., Fukuzaki, N., Kakei, K., Suzuki, T. and Ozeki, S., *Langmuir*, **5**, 960, 1989.
Kaneko, K., *Tanso*, **1993** [No.160], 283, 1993.

Kaneko, K., *Carbon*, **38**, 287, 2000.

Kang, F., Leng, Y. and Zhang, T.-Y., *Carbon*, **35**, 1089, 1997.

Kanoh, H. and Kaneko, K., *J. Phys. Chem.*, **99**, 5746, 1995.

Katawaki, K. and Nishizaki, I., *Pre-stressed Concrete*, **30**, 31, 1988.

Katsuki, H., Natsunaga, K., Egashira, M. and Kawasumi, S., *Carbon*, **9**, 148 1981.

Kimura, S. and Yasuda, E., *Zairyo Kagaku*, **20**, 36, 1983.

Kimura, S., Yasuda, K., Yasuda, E. and Tanabe, Y., *Tanso*, **1987** [No.128], 30, 1987.

Kisamori, S., Kawano, S. and Mochida, I., *Chem. Lett.*, **1993**, 1899, 1993.

Kita, Y., Watanabe, N. and Fujii, Y., *J. Am. Chem. Soc.*, **101**, 3832, 1979.

Kobayashi, K., Miyazaki, K., Ogawa, I., Hagio, T. and Yoshida, H., *Mater. Des.*, **9**, 10, 1988.

Kojima, A., Otani, S., Nakamura, K., Furukawa, S. and Kodaira, I., *Tanso*, **1989** [No.139], 9, 1989.

Kojima, A., Otani, S., Sakaniwa, Y., Furukawa, S. and Kodaira, I., *Tanso*, **1990** [No.141], 23, 1990.

Kojima, A. and Otani, S., *Proceedings of International Workshop on Advanced Materials for Fuctional Manifestation of Frontier and Environmental Consciousness*, Tokyo, Sept. 17–18, 1997.

Konno, H., Nakahashi, T. and Inagaki, M., *Carbon*, **35**, 669, 1997.

Kown, Y.-K., Tomanek, D. and Iijima, S., *Phys. Rev. Lett.*, **82** [No. 7], 1470, 1999.

Koyama, T. and Endo, M., *Ouyou Butsuri*, **42**, 690, 1973.

Koyama, T. and Endo, M., *Jpn. J. Appl. Phys.*, **13**, 1175, 1974.

Kusunoki, M., Rokkaku, M. and Suzuki, T., *Appl. Phys. Lett.*, **71**, 2620, 1997.

Kudryavtsev, Yu. P., Evsyukov, S., Guseva, M., Babaev, V. and Khvostov, V., *"Chemistry and Physics of Carbon"* Vol. **25**, P. A. Thrower, (ed.), p.1, 1997.

Kyotani, T., *Carbon*, **38**, 269, 2000.

Lagrange, P., Metrot, A. and Herold, A., *C. R. Acad. Sci. Paris*, **278**, C-701, 1974.

Lagrange, P. and Herold, A., *C. R. Acad. Sci. Paris*, **281**, C-381, 1975.

Lalancette, J.M. and Roussel, R., *Can. J. Chem.*, **54**, 3541, 1976.

Lavin, J.G., Boyington, D.R., Lahijani, J., Nysten, B. and Issi, J.P., *Carbon*, **31**, 1001, 1993.

Lowell, C.E., *J. Am. Ceram. Soc.*, **50**, 142, 1967.

Mabuchi, A., Tokumitu, K., Fujimoto, H. and Kasuh, T., *J. Electrochem. Soc.*, **142**, 1041, 1995.

Malmstrom, C., Keen, R. and Green, L., *J. Appl. Phys.*, **22**, 593, 1951.

Marsh, H., Heintz, E.A. and Rodriguez-Reinoso, F., *"Introduction to Carbon Technologies"*, Univ. de Alicante, Spain, 1997.

Martens, H.E., Jaffe, L.D. and Jepson, J.E., *Proc. 3rd Conf. on Carbon*, p.529, 1958.

Matsuo, K., Hori, S. and Fukada, K., *Tanso*, **1968** [No.55], 114, 1968.

McKee, D.W., *Carbon*, **24**, 737, 1986.

McKee, D.W., *Carbon*, **26**, 659, 1988.

McKee, D.W., *"Chemistry and Physics of Carbon"*, Vol. **23**, p.173, 1997.

Mering, J. and Maire, J., *"Les Carbones"*, Vol. **1**, GFEC, p.129, 1965.

Meyer, R.A., *Proc. Carbon Conf.*, Baden-Baden, 1986.

Meyer, R.A., *Proc. 19th Conf. Carbon*, p.332, 1989.

Meyer, R.A. and Henderson, G.W., *39th Int. SAMPE Symp.*, p.311, 1994.

Mitsuoka, T., Matsubara, H., Takada, M., Kawamoto, H. and Matsuo, Y., *Trans. Mater. Res. Soc. Jpn*, **14A**, 421, 1994.

Mittal, J. and Inagaki, M., *Synth. Met.*, **92**, 87, 1998a.

Mittal, J. and Inagaki, M., *Synth. Met.* **95**, 21, 1998b.

Mittal, J. and Inagaki, M., *Solid State Ionics*, **121**, 183, 1999a.

Mittal, J. and Inagaki, M., *Synth. Met.*, **99**, 79, 1999b

Miyazaki, K., Kobayashi, K. and Honda, H., *Tanso*, **1977** [No.91], 121, 1977.

Miyazima, T. and Sakai, M., *Fracture Mechanics Ceram.*, **9**, 83, 1992.

Mizutani, Y., Ihara, E, Abe, T., Asano, M., Harada, T., Ogumi, Z. and Inaba, M., *J. Phys. Chem. Solids*, **57**, 799, 1996.

Mizutani, Y., Abe, T., Ikeda, K., Ihara, E., Asano, M., Harada, T., Inaba, M. and Ogumi, Z., *Carbon*, **35**, 61, 1997.

Mochida, I., Korai, Y., Shirahama, M., Kawano, S., Hada, T., Seo. Y., Yoshikawa, M. and Yasutake, A., *Carbon*, **38**, 227, 2000.

Moor, A.W., *"Chemistry and Physics of Carbon"*, Vol. **10**, P.L. Walker, (ed.), New York, p.69, 1973.

Moor, A.W., *"Chemistry and Physics of Carbon"*, Vol. **17**, P.L. Walker, (ed.), p.233, 1982.

Morooka, S. and Kusakabe, K., *MRS Bull.*, **24**, 25, 1999.

Motojima, S., Ueno, S., Hattori, T. and Goto, K., *Appl. Phys. Lett.*, **54**, 1001, 1989a.

Motojima, S., Ueno, S., Hattori, T. and Iwanaga, H., *J. Cryst. Growth*, **96**, 383, 1989b.

Motojima, S., Hasegawa.I., Kagiya, S., Andoh, K. and Iwanaga, H., *Carbon*, **33**, 1167, 1995.

Motojima, S., Iwanaga, H. and Varadan, V.K., *Hyoumen*, **36**, 140, 1998.
Murakami, M., Nishi, N., Nakamura, K., Ehara, J., Okada, H., Kouzaki, T., Watanabe, K., Hoshi, T. and Yoshimura, S., *Carbon*, **30**, 255, 1992.
Murakami, M., Hoshi, T. and Nishiki, N., *Housyako*, **6**, 43, 1993.
Murase, Y., Noda, T. and Inagaki, M., *Tanso*, **1968** [No.54], 80, 1968.
Muto, N., Yanagida, H., Miyayama, M., Nakatsuji, T., Sugita, M. and Ohtsuka, Y., *J. Ceram. Soc. Jpn*, **100**, 585, 1992.
Nagata, Y., Nakama, N. and Saito, K., *Polymer Preprint, Japan*, **40**, 4304, 1991.
Nakagawa, H., *"Shin Tansozairyou Nyuumonn"*, Carbon Society of Japan, (ed.), Realize, Tokyo, p.224, 1996.
Nakajima, T., Kawaguchi, M. and Watanabe, N., *Z. Naturforsch.* **36**, 1419, 1981.
Nakajima, T., Kawaguchi, M. and Watanabe, N., *Carbon*, **20**, 287, 1982.
Nalimova, V.A., Avdeev, V.V. and Semenenko, *Mater. Sci. Forum*, **91–93**, 11, 1992.
Narisawa, M., Adachi, M. and Souma, I., *J. Mater. Sci.*, **29**, 708, 1994.
Nishino, A., *Tanso*, **1988** [No.132], 57, 1988.
Nixon, T.D. and Cawley, J., *J. Am. Ceram. Soc.*, **75**, 703, 1992.
Noda, T., Inagaki, M. and Kato, H., *Bull. Chem. Soc. Jpn*, **35**, 1471, 1962.
Noda, T., Iwatsuki, M. and Inagaki, M., *Tanso*, **1966** [No.47], 14, 1966.
Noda, T., Inagaki, M. and Yamada, S., *J. Non-Cryst. Solids*, **1**, 285, 1969.
Nomine, M. and Bonnetain, L., *J. Chim. Phys.*, **66**, 1731, 1969.
Oberlin, A., Endo, M. and Koyama, T., *J. Cryst. Growth*, **32**, 335, 1976.
Oberlin, A., *Carbon*, **22**, 521, 1984.
Oberlin, A. and Guigon, M., *"Fiber Reinforcements for Composite Materials"*, Chapter 4, A.R. Bunsell, (ed.), Elsevier, Amsterdam, 1988.
Oberlin, A., *"Chemistry and Physics of Carbon"*, Vol. **22**, P.A. Thrower, (ed.), p.1, 1989.
Oberlin, A., Bonnamy, S. and Rouxeht, P.G., *"Chemistry and Physics of Carbon"*, Vol. **26**, P.A. Thrower and L. Radvic, (eds), p.1, 1998.
Ogawa, H., *Nihon Kagaku Kaishi*, **1994**, 560, 1994a.
Ogawa, H., *Nihon Kagaku Kaishi*, **1994**, 809, 1994b.
Ogawa, H. and Saito, K., *Carbon* **33**, 783, 1995.
Ogawa, H., *Carbon*, **38**, 211, 2000.
Ogawa, I., Yoshida, H. and Kobayashi, K., *J. Mater. Sci.*, **16**, 2181, 1981.
Ogawa, I., Kobayashi, K. and Nishikawa, S., *J. Mater. Sci.*, **23**, 1363, 1988.
Okabe, T., Saito, K. and Hokkirigawa, K., *J. Porous Mater.*, **2**, 207, 1996a.
Okabe, T., Saito, K. and Hokkirigawa, K., *J. Porous Mater.*, **2**, 215, 1996b.
Okabe, T., Saito, K., Fushitani, M. and Otsuka, M., *J. Porous Mater.*, **2**, 223, 1996c.
Okada, J. and Takeuchi, Y., *4th Biennial Conference on Carbon*, 1960.
Oshida, K., Ekinaga, N., Endo, M. and Inagaki, M., *Tanso*, **1996** [No.173], 142, 1996.
Otani, S., Phuang, H.L., Kubata, T., Sakaniwa, H. and Suzuki, M., *Denki Kagaku*, **44**, 27, 1976.
Otani, S., *Mol. Cryst. Liq. Cryst.*, **63**, 249, 1981.
Oya, A., *"Shin Tanso Zairyou Nyuumon"*, Carbon Society of Japan, Realize, 1996, p.111.
Oya, A., *"Introduction to Carbon Technologies"*, H. Marsh, E.A. Heintz and F. Rodoriguez-Reinoso, (eds), Univ. de Alicante, p.561, 1997a.
Oya, A., *J. Odor Res. Eng.*, **28**, 57, 1997b.
Ozaki, J., Endo, N., Ohizumi, W., Igarashi, K., Nakahara, M., Oya, A., Yoshida, S. and Iizuka, T., *Carbon*, **35**, 1031, 1997.
Patrick, J.W., (ed.), *"Porosity in Carbons: Characterization and Applications"*, Edward Arnold, London, 1995.
Pfluger, P., Kunzi, H..V. and Guentherodt, H.-J., *Appl. Phys. Lett.*, **35**, 771, 1979.
Pilliere, H., Takahashi, Y., Yoneoka, T., Otosaka, T. and Akuzawa, N., *Synth. Met.*, **59**, 191, 1993.
Plastics and Rubber Institute, ed., *"Carbon Fibers Technology, Uses and Prospects"*, Noyes Publications, New Jersey, 1986.
Pleger, R., Braue, W. and Meyer, R.A., *Proc. 20th Biennial Conf. Carbon*, p.399, 1991.
Rodriguez, N.M., *J. Mater. Res.*, **8**, 3233, 1993.
Rodriguez-Reinoso, F. and Linares-Solano, A., *"Chemistry and Physics of Carbon"*, Vol. **21**, P.A. Thrower, (ed.), p.1, 1995.
Saito, Y. and Uemura, S., *Carbon*, **38**, 169, 2000.
Sakai, M., Miyajima, T. and Inagaki, M., *Composites Sci. Tech.*, **40**, 231, 1991.
Sakai, M., *J. Soc. Mater. Sci., Jpn*, **44**, 138, 1995.
Sanada, Y., Suzuki, M. and Fujimoto, K., *Kasseitann (Activated Carbons)*, Koudannsya Scientific, 1992.
Sato, K., Sato, F. and Tomioka, N., *Tanso*, **1993** [No.157], 107, 1993.
Sato, K., Noguchi, M., Demachi, A., Oki, N. and Endo, M., *Science*, **264**, 256, 1994.

Setoyama, N. and Kaneko, K., *Zairyoukagaku*, **32**, 213, 1995.

Setton, R., *Mater. Sci. Eng.*, **31**, 303, 1977.

Shibata, K., Kasai, K., Okabe, T. and Saito, K., *J. Porous Mater.*, **2**, 287, 1996.

Shiraishi, M., "*Tanso Zairyou Nyuumonn*", Kagakugizyutusha, p.29, 1984.

Sing, K.S., Everett, D.H., Haul, R.A.W., Moscou, L., Pierotti, R.A., Rouquerol, J. and Siemieniewska, T., *Pure Appl.Chem.*, **57**, 603, 1985.

Sogabe, T., Nakajima, K. and Inagaki, M., *J. Mater. Sci.*, **31**, 6469, 1996.

Sogabe, T., Okada, O., Kuroda, K. and Inagaki, M., *Carbon*, **35**, 67, 1997.

Soneda, Y. and Inagaki, M., *Z. Anorg. Allg. Chem.*, **610**, 157, 1992.

Soneda, Y. and Makino, M., *Extended Abstracts of Eurocarbon'98*, Strasbourg, p.827, 1998.

Souma, I., Shioyama, H., Tatsumi, K. and Sawada, Y., *Proc. 4th Jpn. Int. SAMPE Symp.*, Sept. 25–28, 1995.

Strife, J.R. and Sheehan, E., *Ceram. Bull.*, **67**, 369, 1988.

Sugihara, K., Hishiyama, Y. and Ono, A., *Phys. Rev., B*, **34**, 4298, 1986.

Suzuki, T., Kaneko, K. and Gubbins, K.E., *Langmuir*, **13**, 2545, 1997.

Takahashi, Y., *Tanso*, **1993**, 301, 1993.

Tanabe, Y., Yasuda, E., Kimura, S. and Okamine, S., *Tanso*, **1987** [No.131], 181, 1987.

Tanabe, Y., Yasuda, E., Takabatake, M. and Kimura, S., *Report of RLEMIT*, **15**, 83, 1990.

Tanabe, Y., Yasuda, E., Yamaguchi, K., Inagaki, M. and Yamada, Y., *Tanso*, **1991** [No.147], 66, 1991.

Tanabe, Y. and Yasuda, E., *Carbon*, **38**, 329, 2000.

Tanahashi, I., Yoshida, A. and Nishino, A., *Carbon*, **28**, 477, 1990.

Tanaike, O. and Inagaki, M., *Synth. Met.*, **90**, 69, 1997.

Tanaike, O., *Ph.D. Thesis*, Hokkaido Univ., 1998.

Tanaike, O. and Inagaki, M., *Synth. Met.*, **96**, 109, 1998.

Tanaike, O., Hoshino, Y. and Inagaki, M., *Synth. Met.*, **99**, 105, 1999.

Terai, T., *Ph.D. Thesis*, Tokyo Univ., p.206, 1983.

Tibbetts, G.G., *J. Cryst. Growth*, **73**, 431, 1985.

Tibbetts, G.G., Devour, M.G. and Rodda, E.J., *Carbon*, **25**, 367, 1986.

Touhara, H., Kakuno, K. and Watanabe, N., *Tanso*, **1984** [No.117], 98, 1984a.

Touhara, H., Fujimoto, H., Watanabe, N. and Tressaud, A., *Solid State Ionics*, **14**, 163, 1984b.

Touhara, H. and Okino, F., *Carbon*, **38**, 241, 2000.

Touzain, Ph., Michel, J. and Blum, P., *Synth. Met.*, **8**, 313, 1983.

Toyoda, M., Aizawa, J. and Inagaki, M., *Nihon Kagaku Kaishi*, **1998** [No.8], 563, 1998a.

Toyoda, M., Aizawa, J. and Inagaki, M., *Desalination*, **115**, 199 1998b.

Toyoda, M., Moriya, K., Aizawa, J. and Inagaki, M., *Nihon Kagaku Kaishi*, **1999** [No. 3], 193, 1999.

Toyoda, M. and Inagaki, M., *Carbon*, **38**, 199, 2000.

Vignal, V., Konno, H., Inagaki, M., Flandrois, S. and Roux, J.C., *J. Mater. Res.*, **14**, 270, 1999a.

Vignal, V., Morawski, A.W., Konno, H. and Inagaki, M., *J. Mater. Res.*, **14**, 1102, 1999b

Vogel, F.L., *J. Mater. Sci.*, **12**, 982, 1977.

Wang, Z.D. and Inagaki, M., *Tanso*, **1990**, 243, 1990.

Wang, Z.D. and Inagaki, M., *J. Mater. Chem.*, **2**, 629, 1992.

Wang, C.Y., Li, M.W., Wu, Y.L. and Guo, C.T., *Carbon* **36**, 1749, 1998.

Wang, C.Y. and Inagaki, M., *Carbon*, **37**, 147, 1999.

Warren, B.E., *J. Chem. Phys.*, **2**, 551, 1934.

Watt, W. and Johnson, W. *Appl. Polym. Symp.*, No.9, 215, 1969.

Yamada, Y., Imamura, T., Kariyama, H., Honda, H., Oi, S. and Fukuda, K., *Carbon*, **12**, 307, 1974.

Yamaguchi, T., *Carbon*, **1**, 47, 1963a.

Yamaguchi, T., *Carbon*, **1**, 535, 1963b.

Yamamoto, O., Imai, K., Sasamoto, T. and Inagaki, M., *J. Eur. Ceram. Soc.*, **12**, 435, 1993.

Yamamoto, O., Sasamoto, T. and Inagaki, M., *Nihon Seramic Kyokai-Shi*, **102**, 165, 1994.

Yamamoto, O., Sasamoto, T. and Inagaki, M., *Carbon*, **33**, 359, 1995.

Yasuda, E., *Tanso*, **1995** [No. 170], 317, 1995.

Yata, S., Kinoshita, H., Komori, M., Anod, N., Kashiwamura, T., Harada, T., Tanaka, K. and Yamabe, T., *Synth. Met.*, **62**, 153, 1994.

Yazami, R. and Touzain, Ph., *J. Power Sources*, **9**, 365, 1983.

Yoshida, A. and Hishiyama, Y., *Tanso*, **1987** [No.130] 110, 1987.

Yoshida, A., Tanahashi, I., Takauchi, Y. and Nishino, A., *IEEE CHM-10*, **1**, 100, 1987.

Yoshida, A., Tanahashi, I. and Nishino, A., *IEEE CHM-11*, **3**, 318, 1988.

Yoshida, A. and Hishiyama, Y., *Tanso*, **1987** [No. 130], 110, 1987.

Yoshida, A. and Hishiyama, Y., *Tanso*, **1989** [No.137], 93, 1989.

Yoshida, A., Hishiyama, Y. and Inagaki, M., *Carbon*, **28**, 539, 1990.

Yoshida, A., Kaburagi, Y. and Hishiyama, Y., *Carbon*, **29**, 1107, 1991.
Yoshida, A. and Hishiyama, Y., *J. Mater. Res.*, **7**, 1400, 1992.
Yoshida, A., Hishiyama, Y., Ishioka, M. and Inagaki, M., *Tanso*, **1995** [No.168], 169, 1995.
Yoshino, K. and Ueno, H., *Tanso*, **1989**, 29, 1989.

Subject Index